Hans-Jürgen Wiemer/Frans Willem Korthals Altes

Small scale processing of oilfruits and oilseeds

A Publication of
Deutsches Zentrum für Entwicklungstechnologien – GATE
in: Deutsche Gesellschaft für Technische Zusammenarbeit (GTZ) GmbH

Friedr. Vieweg & Sohn   Braunschweig/Wiesbaden

The Authors:
Dr. Hans-Jürgen Wiemer has done his Ph.D. work on socio-economic aspects of rural development at the Technical University of Aachen, Federal Republic of Germany. He is staff member of the Africa-Asien-Bureau, Cologne, where he coordinates projects and studies on agrarian economy and small to medium scale industrialization.
Ir. Frans Willem Korthals Altes is a graduate in chemical engineering from Delft Technical University, The Netherlands. He coordinates the activities in the field of agrotechnology by the rural development programme of the Koninklijk Instituut voor de Tropen, Amsterdam.

CIP-Titelaufnahme der Deutschen Bibliothek

Wiemer, Hans-Jürgen:
Small scale processing of oilfruits and oilseeds / Hans-Jürgen Wiemer ; Frans Willem Korthals Altes. A publ. of Dt. Zentrum für Entwicklungstechnologien – GATE in: Dt. Ges. für Techn. Zusammenarbeit (GTZ) GmbH. – Braunschweig ; Wiesbaden : Vieweg, 1989
ISBN 3-528-02046-6
NE: Korthals Altes, Frans Willem:

The author's opinion does not necessarily represent the view of the publisher.

All rights reserved.
© Deutsche Gesellschaft für Technische Zusammenarbeit (GTZ) GmbH, Eschborn 1989

Published by Friedr. Vieweg & Sohn Verlagsgesellschaft mbH, Braunschweig
Vieweg is a subsidiary company of the Bertelsmann Publishing Group.

Printed in the Federal Republic of Germany by Lengericher Handelsdruckerei, Lengerich

ISBN 3-528-02046-6

Contents

List of Tables 5
List of Flowsheets 5
List of Figures 5
Preface 7
0. Introduction 8
   0.1 Economic aspects 8
   0.2 Technical aspects 12
      0.2.1 Processes for oil fruits 13
      0.2.2 Processes for oil seeds 14
   0.3 Development potentials 15
1. Oilplants and their Potential Use 18
   1.1 Characteristics of vegetable fats and oils 18
   1.2 The major oilplants 19
      1.2.1 Oil palm 19
      1.2.2 Coconut palm 20
      1.2.3 Soyabean 21
      1.2.4 Groundnut 23
      1.2.5 Sunflower 24
      1.2.6 Sesame 25
      1.2.7 Rape and mustardseed 26
      1.2.8 Other oil-yielding plants 28
   1.3 By-products 30
   1.4 Further processing 32
2. Target Groups and Technologies 34
   2.1 Family level 34
      2.1.1 Oil palm fruit 34
      2.1.2 Oil seeds 36
   2.2 Village level 39
      2.2.1 Oil palm fruit 39
      2.2.2 Oil seeds 44
   2.3 District level 52
3. Case Studies 54
   3.1 Shea nut processing by women in Mali 54
   3.2 Hand-operated sunflowerseed processing in Zambia 57
   3.3 Oil palm fruit processing as a women's activity in Togo . . 61

4. Financial Analysis of the Case Studies 64
   4.1 Shea nut processing in Mali 65
   4.2 Sunflower seed processing in Zambia 69
   4.3 Oil palm fruit processing in Togo 74
5. Selected Equipment 78
   5.1 Hand-operated equipment 78
      5.1.1 Hand-operated processing of oil palm fruit 78
      5.1.2 Hand-operated processing of oil seeds 81
   5.2 Motorized equipment 85
      5.2.1 Motorized processing of oil palm fruit 85
      5.2.2 Motorized processing of oil seeds 87
6. Ongoing Research and Development Work 92

Annex
1. Guidelines for the preparation of oilfruit or oilseed processing projects . . 94
2. List of abbreviations and addresses 94
3. Calculation of internal rate of return (IRR) 97
4. Summary of hand-operated processes 98
5. Currency conversion table 99
6. Literature 99

List of Tables

1 World Production of Major Oil Seeds 8
2 World Production of Major Vegetable Oils 9
3 World Exports of Major Oil Seeds 10
4 World Exports of Major Vegetable Oils 10
5 World Exports of Oilseed Cake and Meal 11
6 Important Oilcrops and their By-Products 30
7 Examples of the Chemical Composition of Oil Cakes fit for Animal Feed . 32
8 Palm Oil: Oil Recoveries Obtained with Different Processes and Equipment 44
9 Typical Performance of KIT/UNATA Hand-operated Equipment 48
10 Typical Performance of IPI Hand-operated System 50
11 Typical Performance of Oil Expeller MINI 40 51
12 Typical Performance of Oil Expeller MRN (AP VII) 51
13 Assumptions for Shea Nut Processing in Mali 65
14 Assumptions for Sunflower Seed Processing in Zambia 70
15 Assumptions for Oil Palm Fruit Processing in Togo 75
16 Weight and Capacity of IPI Equipment 84
17 Weight and Capacity of UNATA Equipment 85

List of Flowsheets

1 Traditional Process for Oil Palm Fruit 35
2 Traditional (wet) Process for Processing Oil Seeds 37
3 Usual Process for Oil Palm Fruit with Hand Press 41
4 KIT Process for Oil Palm Fruit with Hand Press 42
5 Dry Process for Processing Oilseeds 47

List of Figures

1 World Prices for Selected Vegetable Oils 12
2 Oil palm 19
3 Coconut Palm 20
4 Soyabean 22
5 Groundnut 23
6 Sunflower 25
7 Sesame 26
8 Mustard 27
9 Duchscher Curb Press 40
10 Palm Fruit Pounder (TCC) 43
11 CALTECH Oil Press (APICA) 44
12 Power Ghani 45
13 Palm Nut Cracker (KIT/UNATA) 46
14 Cocos Grater (KIT) 46

15	Roller Mill (KIT/UNATA)	46
16	Heating Oven (KIT)	48
17	Spindle Press (UNATA)	49
18	Oil Yield as Determined by the Number of Machines Used in the Process for Sunflower Seeds (IPI)	50
19	Shea Nut Processing in GTZ/GATE Project	55
20	Sunflower Seed Processing in Zambia	59
21	Oil Palm Fruit Processing in Togo	62
22	Summary Sheet, Shea Nut Processing in Mali	68
23	Summary Sheet, Sunflower Seed Processing in Zambia	72
24	Summary Sheet, Oil Palm Fruit Processing in Togo	76
25	Oil Press, as disseminated by ENDA	78
26	TCC Press	79
27	Hydraulic Press, Usine de Wecker, Typ 50—83	81
28	8 L Spindle Press, KIT/TOOL and Kettles	81
29	Shea Nut Press, GTZ/GATE	82
30	IPI Decorticator/Winnower	83
31	IPI Seed Crusher (Roller Mill)	83
32	IPI Seed Scorcher	83
33	IPI Oil Press	84
34	Speichim Expeller Press	86
35	Cecoco Press, Hander 52	87
36	Simon Rosedowns, Press MINI 40	88
37	Montforts + Reiners, „Komet" Double Screw Expeller	89
38	Reinartz Screw Press AP VII	90

Preface

Processes for the extraction of vegetable oil from oilfruits and oilseeds are known in many countries. The technologies for these processes are, in most cases, either still traditional or very modern. Traditional technologies usually have the advantage of requiring low investments, but are labour-intensive and time-consuming. Sophisticated large scale technologies, on the other side, are generally beyond the financial reach of the rural population in developing countries.

The present publication aims at closing an information gap on a third option: small scale or intermediate technologies for oil extraction. These technologies have been developed by various institutions and are presented in the hope that they might contribute to more appropriate solutions and generate additional income for families, self-help groups and co-operatives, particularly in rural Africa.

The German Appropriate Technology Exchange (GATE), as a department of the Deutsche Gesellschaft für Technische Zusammenarbeit (GTZ) GmbH, has gained project experience with women groups in West Africa using small scale equipment. The experience includes that, apart from technical aspects, the social setting and the economic effects have to be considered before the introduction of new technologies. GTZ/GATE has therefore asked the AFRICA ASIEN BUREAU, Cologne, and the KONINKLIJK INSTITUUT VOOR DE TROPEN, Amsterdam, to combine their expertise for this publication.

The authors sincerely hope that the language they have used is not too technical or complicated for those readers who are active in rural development, but have only a basic understanding of the agronomic, technical and financial implications of oil processing.

All those, who have contributed to the realization of this publication, are greatfully acknowledged, in particular Mr. R.D. Heubers and Mr. R. Merx, staffmembers agrotechnology of KIT and Mr. G. Espig of the University of Stuttgart-Hohenheim.

0. Introduction

0.1 Economic aspects

Over the last two decades, world production and consumption of oil fruits and oil seeds and their products has almost doubled. In terms of value, oil fruits and oil seeds now take third place after starch plants and fruits/vegetables.

Apart from their nutritious value, oil fruits and oil seeds are of particular economic importance for the developing countries, from where a substantial part of the (statistically registered) world production originates. In Africa, Asia and Latin America, the cultivation of oil plants not only plays a major role in the provision of protein and fats, but contributes considerably to exports earnings.

Table 1 reflects the development of world production of major oil seeds as registered by the Food and Agricultural Organization of the United Nations. The figures given for the last 15 years, however, do not necessarily reflect the total for each crop, since, in many countries, there is a significant production for the growers' domestic requirements which never appears in official statistics.

Table 1: **World Production of Major Oil Seeds**
(million tons)

	1969/71	1979/81	1983	1984	1985
Soyabeans	43.5	86.0	79.4	90.2	100.1
Coconuts	29.4	33.2	34.2	33.1	34.7
Cottonseed	22.7	27.3	27.6	35.1	32.2
Groundnuts	17.9	18.6	19.0	20.3	21.3
Sunflower Seed	9.9	14.4	15.6	16.4	19.1
Rapeseed	6.6	11.2	14.0	16.6	18.9
Palm Kernels	1.2	1.8	2.1	2.4	2.7
Linseed	3.5	2.5	2.3	2.5	2.5
Sesame Seed	1.9	2.0	2.1	2.0	2.4
Safflower Seed	0.7	1.0	0.9	0.9	0.8
Castor Beans	0.8	0.8	0.9	1.0	1.3

Source: FAO Production Yearbooks, Vols. 35–39
- For Coconuts, production is expressed in terms of weight of the whole nuts, excluding only the fibrous outer husk, whether ripe or unripe, whether consumed fresh or processed into copra or dessicated coconut
- For Cottonseed, direct production figures are reported by countries accounting for about 60 % of world production; data for the remainder are derived from ginned cotton production
- Groundnuts in shell
- For Rapeseed, figures include mustard seed for a few countries (e.g. India and Pakistan)

Table 2: **World Production of Major Vegetable Oils**
(million tons)

	1972/73	1976/77	1982/83	1984/85	1985/86
Soyabean Oil	7.41	9.95	13.57	13.25	13.53
Palm Oil	1.83	2.91	5.91	6.94	7.74
Sunflowerseed Oil	3.31	3.39	5.60	6.06	6.49
Rape/Mustard Oil	2.49	2.58	4.98	5.69	6.00
Cottonseed Oil	2.89	2.70	3.14	3.84	3.64
Groundnut Oil	2.50	2.92	2.86	3.10	3.20
Coconut Oil	2.49	2.70	2.76	2.93	3.09
Olive Oil	1.55	1.52	1.91	1.63	1.24
Palm-Kernel Oil	0.42	0.53	0.78	0.92	1.06
Linseed Oil	0.66	0.64	0.64	0.66	0.65

Sources: Oil World, Vol. XX. No. 13, p. 302 f
Marchés Tropicaux, No. 2112, p. 1167

As can be seen in Table 2, soyabeans have established a leading position and now represent more than 40 % in volume terms of world production. If the annual increase in production is considered, rapeseed, with an average increase of more than 12 % per year, has an even more positive trend than soyabeans, followed by palm kernels and sunflower seed. Palm kernels are, in fact, to be seen as by-product of palm oil production. The production of palm oil fruit, however, is not recorded, and therefore palm kernels can indicate the growth for the whole crop.

World production of major vegetable oils, as shown in Table 2, has been steadily increasing over the past decades with average annual growth rates of more than 5 %. In volume terms, the total (recorded) production now stands at more than 60 million tons per year, which contributes more than two thirds to the world consumption of all edible oils.

The great increase in the production of soyabean oil has been the major development on this market in the past decades. Due to the introduction of the hydrogenation technique, which made the oil suitable for margerine, and the rapid increase in demand for soya cattle cakes, world production of soyabean oil tripled over the last 20 years.

The production of vegetable oils from palmfruits, sunflowerseed and rape/mustard has doubled over the same period. The group of oils made from cottonseed, groundnut and coconut, however, is more or less constant. Since many vegetable oils are direct competitors, the relative importance of these oils may well change in the future, but the dominance of soyabean oil is unlikely to change.

In terms of volume of world production, other vegetable oils are of minor or only regional importance. Olive oil, for example, is almost exclusively produced in the mediterranean countries where it meets a unique consumer preference. Since it has not gained acceptance in other regions, olive oil will not be dealt with in the present publication. Other oils, however, are of particular relevance for specific areas in developing countries (e.g. sheanut in West-Africa) and will therefore be described in more detail than their overall importance would appear to justify.

Table 3: **World Exports of Major Oil Seeds**
(million tons)

	1970	1975	1980	1984	1985
Soyabeans	12.62	16.56	26.88	25.78	25.53
Rape/Mustard	1.23	1.05	2.07	2.79	3.64
Sunflowerseed	0.48	0.35	1.94	2.15	1.78
Groundnuts	0.93	0.82	0.74	0.75	0.80
Linseed	0.63	0.36	0.53	0.55	0.70
Copra	0.92	1.09	0.44	0.29	0.38
Sesameseed	0.22	0.21	0.21	0.31	0.30
Cottonseed	0.49	0.20	0.32	0.21	0.26
Palm Kernels	0.46	0.34	0.20	0.13	0.09
Castor Beans	0.13	0.09	0.06	0.11	0.09

Source: FAO Trade Yearbooks, Vols. 25–39
– For Groundnuts, nuts reported in the shell are converted to shelled equivalent using a conversion factor of 70 %
– For Rape and Mustard seed, most trade statistics do not allow a distinction, and figures in many instances even include other minor oil seeds

World trade in oil fruits and oil seeds, as shown in Table 3, is even more dominated by soyabeans than the production figures seem to indicate. With one quarter to one third of the total production entering the world market, soyabeans is in fact the only oilcrop with considerable exports in an unprocessed condition. Most prominent exporter is the USA, which has two thirds of the whole market. The soyabean

Table 4: **World Exports of Major Vegetable Oils**
(million tons)   * no records

	1970	1975	1980	1984	1985
Palm Oil	0.91	2.05	3.59	4.30	5.23
Soyabean Oil	1.12	1.36	3.20	4.03	3.49
Sunflower Oil	0.73	0.79	1.15	1.67	1.94
Rape/Mustard Oil	0.18	0.35	0.69	1.07	1.32
Coconut Oil	0.62	1.03	1.21	0.99	1.23
Palm Kernel Oil	0.16	0.26	0.38	0.54	0.65
Olive Oil	0.26	0.19	0.30	0.36	0.50
Cottonseed Oil	0.25	0.37	0.44	0.34	0.39
Groundnut Oil	0.43	0.40	0.49	0.29	0.32
Maize (Corn) Oil	*	*	0.21	0.31	0.30
Linseed Oil	0.27	0.20	0.34	0.29	0.27
Castor Oil	0.19	0.19	0.17	0.18	0.18

Source: FAO Trade Yearbooks, Vols. 25–39
– In general, figures do not reflect entire world trade since some national statistics classify all or a large part of their trade under such headings as „edible oils", „vegetable oils, nes" etc. and therefore have been omitted
– For Palm oil, data refer to trade in crude and refined oil as well as trade in palm oleine and palm stearin

exports of all developing countries together only amount to about half of this share, showing South America (Argentina and Brazil) in a leading position. The main direction of the soyabean trade is towards Europe, which imports more than half of the available quantity.

Of more relevance to the developing countries is the world trade in vegetable oils. As shown in Table 4, palm oil alone accounts for about one third of the trade, the major exporters being Malaysia and Indonesia, with Singapore as the major port of the region. Prior to 1945, Africa produced and exported most palm oil, but has now lost this position due to the massive planting of high yielding varieties in South East Asia. For palm oil, Asia has also the highest demand, although import statistics do not correspond to actual consumption due to substantial re-exports (Singapore).

Whereas coconut and palm oils are almost exclusively exported from developing countries, they are able to supply three quarters of the world market for linseed oil and groundnut oil and even realize shares of about 50 % for sunflower, cottonseed and soyabean oils. Major suppliers in these markets are currently Malaysia (palm and palm kernel oil), Philippines (coconut oil), Brazil (soyabean and groundnut oil) and Argentina (linseed and sunflower oil).

The world trade in vegetable oils is, to a considerable extent influenced by the demand for by-products (oilseed cake and meal) and the prices of others chemically similar, i.e. competing vegetable oils.

As shown in Table 5, the trade in oilseed cake and meal consists to more than 70 % of the by-product of soyabean; a crop for which the meal is – due to the low oil content – relatively more important than the oil. A great advantage of the soyabean and, in fact, a reason for the production increase, is that it can be freely fed to all livestock groups. Major exporters of soyabean meal are Brazil, USA and Argentina, the major importing region is West Europe.

As illustrated in Figure 1, world market prices for vegetable oils depend as much on the specific variety of oil as on apparently periodical fluctuations. The different characteristics of vegetable oils, which

Table 5: **World Exports of Oilseed Cake and Meal**
(million tons)

	1970	1975	1980	1984	1985
all cakes/meals	11.1	13.8	25.7	28.5	30.5
– of soyabean	5.4	8.7	17.9	20.3	21.9
– of sunflower	0.6	0.4	0.9	1.3	1.7
– of rapeseed	0.2	0.2	0.6	1.3	1.4
– of palm kernels	0.2	0.4	0.5	0.7	1.0
– of cottonseed	1.4	1.1	0.8	0.5	0.7
– of copra	0.6	0.7	1.1	1.0	0.7
– of linseed	0.6	0.4	0.7	0.7	0.6
– of groundnuts	1.5	1.2	1.1	0.6	0.4
– of other seeds	0.6	0.6	2.1	2.5	1.9

Source: FAO Trade Yearbooks, Vols. 25–39

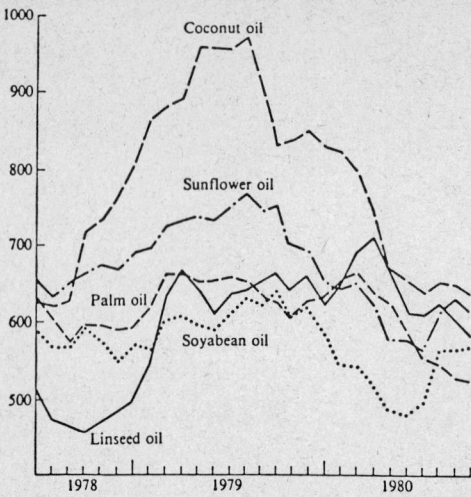

Figure 1: World Prices for Selected Vegetable Oils (monthly average quotations, West European ports) in US $ per ton. Source: E. A. Weiss, Oilseed Crops, 1983, p. 9

determine the relative scarcity and the possible use of the product, will be discussed in more detail in Chapter 1.1 of this publication.

The price fluctuations tend to assume a cyclic pattern with a periodicity of several years. During this period, stock build-up leads to more demand and subsequently higher prices, followed by a period of stock liquidation with the opposite price effect with shorter periods of market consolidation in between. As reported in market observations of „Marchés Tropicaux" (5/86), prices for specific oils have, over the last few years, shown fluctuations of more than 300 % (based on the lower price). In absolute terms, the price for coconut oil, for example, rose from about 400 $/t (CIF, Rotterdam) in 1983 to about 1400 $/t in the following year only to fall back to the 1983 low the year after. Except for soyabean, prices in the last two to three years have taken another drastic downturn and are likely to stabilize at a relatively low level.

For certain vegetable oils or specific qualities of oils, only regional or even local markets have been established. Accordingly, no representative market data are available for these products. For the African context, however, the following examples should demonstrate that these local markets can show particular consumer preferences and do not necessarily follow the same trend as the world market for the major varieties.

In nearly all oil palm growing countries in West Africa, the red palm oil is an important food ingredient. The colour and the taste from the oil of the traditional dura variety are generally more highly valued than those from the (hybrid) tenera variety. Also, the taste of the oil processed by traditional methods is preferred. In Benin, Cameroon and Nigeria about 50 % of the palm oil consumed is produced in the traditional way. In Ghana and Sierra Leone it is even 70 % and 90 % respectively.

In the Pacific Islands and along the coast of West Africa, coconut oil is also produced in a traditional way. Although this kind of oil has a particularly preferred taste its keeping quality is reported to be low. Shea nut butter is an important fat in Burkina Faso and Mali and also has a special taste that is highly regarded.

Most of these traditionally produced oil varieties are only available in a certain season of the year, which leads to low market prices at harvest time and conversely to high prices out of season.

0.2 Technical aspects

The present sub-chapter aims at giving an outline of the existing processes for manufacturing vegetable oils without entering into complex technical details.

In the presentation, an initial distinction is made between the aspects involved in the processing of oil fruits and those involved in the processing of oil seeds. For oil fruits, the processes for the extraction of oil from oilpalmfruit is given as the most important example. For oil seeds, the principles for different crops are sketched.

In oilcrop processing, many technologies have been developed, which have their place in different economic and cultural situations. Therefore, a second distinction is made between traditional methods and modern methods in each of the above categories. Whereas traditional methods are seen as clearly reflecting their social environment, modern industrial processes are the result of high level technological experience.

Small scale oil production systems, as the main target of this publication, try to combine these two characteristics; i.e. they should be adaptable to a given social context, and they should be technically efficient and reliable (which in any case requires proper maintanance and an adequate supply of spare parts). Small scale systems, therefore, can only be improved on the basis of an understanding of traditional methods and a thorough knowledge of modern technology. The presentation of this intermediary technology level, however, is given in more detail in later chapters.

0.2.1 Processes for oil fruits

Since the oilpalm gives the economically most important tropical oilfruit, the technologies for its extraction can serve as an example in this category.

In the *traditional process*, the fruit is first removed from the bunches, generally after the bunches have fermented for a few days. The fruit is then cooked and pounded or trampled. The mashed mass is mixed into water. The oil and oil-containing cell material is separated from the fibre and the nuts by rinsing with excess water and pressing by hand. The oil-containing mass, now floating on the top, is collected and boiled. In this step, the oil separates from the rest and collects on the surface. It is skimmed off and finally dried.

The actual execution of the process may vary somewhat from area to area; most traditional processes, however, have in common the superfluous use of water. Using this process, generally not more than 50 % of the oil is obtained. The problems are:
– the digestion by means of pounding or trampling,
– the separation of the oil and oil-containing material from the fibres and the nuts by means of water and
– the liberation of the oil by cooking afterwards.

The potential for improvement of this technology and thereby the development of small scale extraction equipment in principal depends on
– better cooking by means of steam,
– better digesting using a reheating step with steam and
– more effective pressing in a batch press or continuously working screw press.

The *modern process* of extracting palm oil, used on a larger scale, starts with the steam sterilization of the bunches. The bunches are threshed and the fruit is digested mechanically, while heated with steam. The mass is then pressed in hydraulic presses or continuously in screw presses. The oil is separated from the press fluid by heating and is finally dried.

0.2.2 Processes for oil seeds

In addition to the distinction made between traditional and modern methods, the processes for oil seeds should also be divided into so-called wet and dry extraction methods.

Of the *traditional wet processes*, the extraction of coconut oil from fresh coconuts is the best known. It starts with grating the meat, after which the oil as well as the proteins and impurities are extracted as a milk from the fibrous residue by pressing (by hand or foot) and rinsing with fresh water. The milk is left to stand to form an oilrich cream on top. The cream is boiled to separate the oil from water and other impurities. The oil can be skimmed off. It still contains a protein- rich residue that can be filtered off after drying and used for human consumption.

Other oil seeds, like groundnuts, palmkernels and sheanuts are roasted and crushed as fine as possible (e.g. first by pounding, followed by crushing between stones or a stone and an iron bar). The crushed mass is mixed with water, and the oil is obtained by cooking the mixture, causing the oil to float. The oil is finally skimmed off and dried by heating. Sheanut oil is often obtained by beating air into a mixture of crushed seeds with some water using a hand-operated buttermaking process. The milk or cream floating on top of the beaten mass at the end of the process is then cooked to evaporate the water and dry the oil.

The weak points of these processes are the grating or crushing steps. They are time-consuming and exhausting work, yet crushing is generally not fine enough. Thorough crushing can improve the oil recovery considerably. In many areas, engine-driven discmills are used by women in small commercial enterprises to get their seed crushed.

All the *traditional dry processes,* as well as the modern dry-extraction methods, consist of three essential unit operations:
– size reduction,
– conditioning by heating and
– eparation of the oil.

The difference between the dry processes is the way by which the oil is separated. With respect to this difference, the following traditional methods can be distinguished:

– without pressure,
– with a wedge press,
– with a screw press,
– with a beam press or
– with a ghani (which combines the above unit operations).

In historical perspective, the use of pressure in the process seems to indicate a society with a higher technical level of achievement. Wedge, beam and screw presses have already been used by the Egyptians, Romans and Chinese. The beam press, which took up a lot of room, was soon superseded by wedge and screw presses, which work fairly satisfactorily. The animal-driven ghani is mainly used on the Indian sub-continent, from where it originates.

Traditional dry processes are very labour intensive and improvements seem appropriate, at least for any kind of market-oriented production. The potential for improvements would best be tapped by the introduction of simplified versions of the modern technologies (see below); e.g. by

– crushing the seed in a roller mill,
– „cooking" in a directly fired pan,
– pressing with an unsophisticated spindle press, a hydraulic press or an engine-driven oil expeller.

In India, the productivity of ghanis has been drastically improved by the introduction of motor-driven versions, which are fast replacing the animal-driven ones.

As mentioned, the *modern dry processes* consist of the same unit operations as the traditional extraction methods. First, the shells or hulls are separated from the nut- or seed kernels to obtain a mass with a maximum oil content. Palmnuts from the African oilpalm or American palms are cracked. Seeds, such as groundnut, sunflower, cotton and kapok are decorticated. The oil-containing kernel material is then milled between rollers to obtain a well-crushed material in the form of flakes. The crushed mass is „cooked" in a set of steam heated pans in a humid atmosphere and subsequently dried.

The dry mass is then pressed, a process which generally is applied twice; i.e. pre-pressing and deep-pressing. The extraction generally takes place by means of oil expellers. Finally, the oil is filtered.

The modern processes, as opposed to the traditional methods,

- apply higher pressures and gain higher yields,
- use power-driven size-reduction equipment and therefore have a higher power consumption per kg of oil,
- are less labour intensive, but require higher initial investments,
- involve less variable, but more constant costs.

The most modern process is the solvent extraction. In this process, the reduced seeds are chemically extracted with a non-polar solvent (usually hexane). In contrast to the modern dry processes with expellers, the solvent extraction cannot be carried out economically on a small scale.

Nevertheless, the process has a number of advantages, such as

- high extraction yield (95 – 99 %),
- high capacity continuous process.

For the purpose of a small-scale production, the main disadvantages, such as:

- large initial investment capital needed,
- large maintenance costs and
- need for skilled labour

are decisive, in particular for most projects in developing countries.

Modern wet processes have been developed for coconut and groundnut processing, but are not likely to become economic. Therefore, the modern wet processes will not be further discussed in this publication.

0.3 Development potentials

The concept of rural development aims to meet the basic needs of the majority of the population in Africa, Asia and Latin-America. This direct approach to basic needs has taken a variety of forms, project types and degrees of integration of individual measures, ranging from infrastructure projects (e.g. transport, energy and water supply), institution building (agricultural and other extension services, assistance to self-help groups and cooperatives), supply with credits and inputs and the promotion of local, small-scale processing of agricultural products.

Agro-industrial development, defined as processing of agricultural products on a larger scale, is of necessity, often located in central places of the producing areas or the capital of the country. The basic concept is rather a macro-economic approach of substituting imports of consumer goods or earning foreign currency with exports. Agro-industrial projects, such as the establishment of central oil mills, have often been successful in keeping an additional value-adding processing step in the hands of a developing country and have thereby led to a more advantageous participation in the structure of post-colonial trade.

The other side of the coin is, however, that quite a number of these projects have not proved to be viable due to insufficient or badly planned rawmaterial supply, management problems and/or highly fluctuating market prices. In macro-economic terms, the result of such projects has often been an increase in foreign debts rather than any profit for the country.

Without unfair generalization, one might say that very few agro-industrial plants have made a substantial direct contribution to the basic needs of rural people. As far as oil mills are concerned, producer prices are often kept artificially low in order to be competitive in international markets. Additional employment opportunities are relatively few, and the improved supply of vegetable oil is more often geared to the urban population. Producers of vegetable oil in rural areas, especially women, often find it more difficult to compete in local markets, since their traditional techniques are very labour-intensive and relatively inefficient. Provided that the social context is considered, as sketched below, the promotion of improved small-scale equipment for oil processing could therefore close a technological gap, increase the availability of oil for personal consumption and generate income in rural areas.

All societies in developing countries are characterized by a specific division of tasks between men and women. Not only in developing countries, but worldwide, the tasks of childrearing and housekeeping are attributed to the female members of the economic unit; i.e. the nuclear family, the extended family, the tribe. etc.

In most developing countries, especially in Africa, housekeeping comprises all family-related activities, often without the possibilities available in industrialized countries to make use of external services and institutions. In rural areas, housekeeping includes a broad spectrum of time consuming tasks, varying from the provision of water and fuel wood, the preparation of meals, washing, cleaning, most handicraft, the cultivation of vegetable gardens to the processing of basic food (such as the production of vegetable oil).

In addition to childrearing, which is a particular stress situation with every additional baby, the numerous tasks of women in rural Africa amount to average work loads of 16 hours a day. This is a considerably heavier burden than any man would normally carry, and is, in itself, a convincing argument for regarding rural women as a target group which deserves particular development efforts.

The disadvantageous division of responsibilities is, however, not limited to the work load as such: In most African countries, rural women have to take care for at least a part of the financial needs for housekeeping (for food, clothing, medicine, schoolbooks, etc.). For this purpose, cash crops have to be cultivated and marketed, handicraft articles produced and sold and other services provided. In the West African context, the production of vegetable oil plays an important role.

The necessity for independent financial resources for women also stems from the fact that men often consider any additional income should be for the head of the household as a contribution to his personal consumption (for radios, bicycles, alcohol, etc.). Furthermore, migration away from rural areas in Africa has already lead to 30 % of female heads of households, who are more or less solely responsible for all needs of the family.

The improvement of traditional techniques and the transfer of appropriate technology (small scale) for the local production of vegetable oil can therefore be seen as a contribution to improve the social and economic living conditions of rural women. Depending on the social context, which varies widely, the introduction of new processes might, however, also face problems. Difficulties might arise from:
- the access to sources of finance, in particular credit, which – for one reason or another – is more freely given to men,
- the volume of the necessary investment, in particular for power assisted technology, which poses a considerable risk and is often intimidating to women,
- the financial needs for production costs, in particular fuel for motor-driven versions which may not always be readily available in rural areas,
- the dependence on repair and maintenance services by workshops in the village or even the next city,
- the need for a minimum degree of organization beyond the family level; i.e. in self-help groups, informal pre-cooperatives or even cooperatives with statutes and formal membership,
- the danger of men taking over after the successful introduction of the new and attractive technology, either for reasons of prestige or as a source of income.

The above mentioned potential difficulties might not be valid in specific cultural settings; in others, even one of these points could well lead to a complete failure of a project. In particular the last two points emphasize the necessity of a detailed knowledge of the social background at the village and even family level before starting to promote new technologies for local oil processing.

In West Africa, project experience has shown rather stable structures of women self-help groups. The transfer of this experience to, for example, East African countries should, however, be handled with some caution. The traditionally less autonomous status of women in this region might make more formal structures of organization (cooperatives) necessary.

A simple transfer of appropriate technology, therefore, appears to be insufficient to reach rural women as target groups of development efforts. Rather, a social approach has to be chosen, which starts with a careful identification of existing forms of organization, includes a training component to strengthen these structures and thereby develops and secures independent sources of income for rural women.

Although the purpose of the present publication basically is to provide technical information, the social approach — after the characterization of the major oilcrops and vegetable oils in Chapter 1 — is reflected in the main parts of the booklet. Chapter 2 first identifies socio-economic units (e.g. family, village, district), then describes the technology which could be considered for each of these units. Chapter 3 gives examples from project experience introducing improved technology at the village level. In Chapter 4, the economics for the case studies are analyzed and alternative technical solutions evaluated. Finally, the concluding chapters provide technical details, addresses of institutions and companies, a look at current research, guidelines for the identification of an oil processing project and a short list of relevant literature.

1. Oil Plants and their Potential Use

1.1 Characteristics of vegetable fats and oils

In principle, there are no essential differences between vegetable fats and oils. The distinction is only a question of melting points, fats being solid and oils being liquid at the temperature concerned.

Chemically, fats and oils are glycerides. A glyceride is a combination of glycerol with fatty acids, a so-called ester. This compound can be split up by naturally occurring enzymes, which are generally present in the rawmaterial, and by alkali. The latter reaction is essential for the production of soap. In the case of enzymes, free glycerol and free fatty acids are formed, a process that also takes place when fats are digested in the human body.

The fatty acids found in vegetable fats and oils are generally based on 12 to 20 carbon atoms. They can be saturated or unsaturated. Saturated fatty acids contain only carbon atoms linked to not less than two hydrogen atoms; unsaturated fatty acids contain atoms with fewer hydrogen atoms, resulting in so-called double bonds.

The more common saturated fatty acids are referred to by name, e.g.:
- lauric   (C12),
- myristic   (C14),
- palmitic   (C16),
- stearic   (C18),
- arachidic   (C20).

The same applies for unsaturated acids, e.g.:
- oleic (C18) with one double bond (9:10),
- linoleic (C18) with two double bonds (9:10, 12:13),
- linolenic (C18) with three double bonds (9:10, 12:13, 15:16).

Vegetable fats and oils have high calorific values and are therefore important sources of energy for the human diet. Besides, they contain so-called „essential" fatty acids (i.e. those necessary for good health) which animals cannot synthesize. Vegetable fats and oils also serve as carriers of the fat soluble vitamins, such as A, its provitamine Carotene, D, E (tocopherol) and K. Furthermore, fats and oils are, of course, important in giving taste to the food.

Fats and oils are relatively stable products. However, the quality of the fats or oils can be harmed by reactions which cause the formation of free fatty acids or rancidity. These reactions are caused by enzymes, air or moulds (so- called ketone rancidity). Fats can be split by active enzymes if the required reaction conditions are fulfilled (high temperature and high moisture content).

To prevent enzymatic reactions, oxidation and/or mould growth, vegetable oils and fats should be stored:
- at a relatively low temperature,
- airtight,
- dry,
- clean and
- in the dark.

Proper storage can be in dry, clean containers such as bottles, tins or drums, filled to the top and well closed. To prevent oxidation, the oil should contain an antioxidant such as tocopherol (vitamin E). As mentioned above, tocopherol is to some degree – depending on the nature of the raw material – already present in unrefined oils and, thus, acts as a natural antioxidant.

When stored in this way, vegetable oils and fats have a „shelf life" (remain fresh) for at least six months.

1.2 The major oil plants

1.2.1 Oil palm

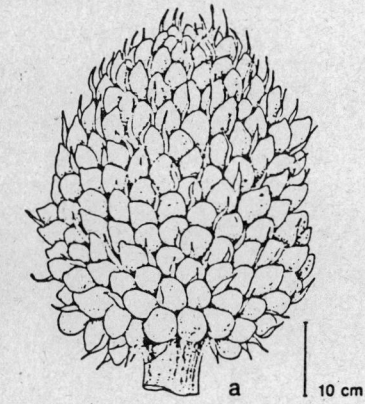

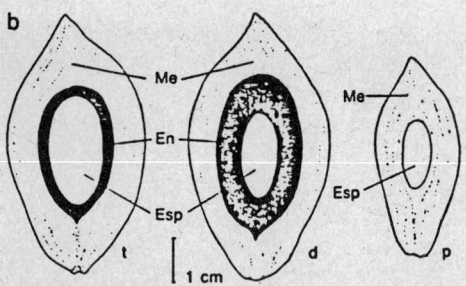

Figure 2: Oil palm
a) bunch, b) fruit of tenera (t), dura (d) and pisifera (p), Me = mesocarp, En = endocarp, Esp = endosperm. Source: S. Rehm, G. Espig, 1984, p. 83

Botanically, oil palms are groups in the genus Elaeis, of which the major varieties are the African oilpalm *(E. guineesis)* and the American oilpalm *(E. oleifera)*. Both can easily be crossed and give fertile hybrids.

Oil palms need an even temperature of between 24° C and 28° C; a reason why the cultivation is limited to the wet tropics about 10° north and south of the equator and below altitudes of 50° metres above sea level. Favourable conditions are annual rainfalls of 1500 to 3000 mm with dry seasons not exceeding three months.

Apart from tropical West-Africa, oil palms are mainly cultivated in South-East Asia and, as a later development, in Central and South America. Due to the introduction of new varieties, world production of oil palm fruit has at least doubled in the last decade as indicated by the growth in the production of kernels to 2.7 million tons in 1985 (see Table 1). Major producing countries of palm kernels are Malaysia (1.2 million tons), Nigeria (0.4 million tons) and Brazil (0.3 million tons). The world market for palm oil is dominated by Malaysia, Singapore and Indonesia, which have taken a market share of over 90 %; Africa as a whole only contributes 2 %.

Oil palms in a good growing condition carry in each leaf axil a flower from which a bunch (see Figure 2) with 1000 to 4000 egg-like, 3 to 5 cm long fruits can develop.

The majority of the oil is contained in the flesh of the fruit (mesocarp), about 12 % is in the inner nut (endosperm). According to the thickness of the nut shell (endocarp), three types are distinguished: *dura* with a shell thickness of 2 to 8 mm, *tenera* with 0.5 to 3 mm and *pisifera* without shell.

The selection of high-yielding types and the breeding of hybrid seeds (tenera is a cross between dura and pisifera) have increased the average yields of oil palms enormously. Wild growing bush palms in Africa used to give 0.6 tons/ha of oil; current yields are more than 6 tons/ha and make the oil palm the highest-yielding oil plant. Hybrid seeds for tenera varieties, however, have to be produced in specialized seed centres. Aims of the breeders are, apart from a thin endocarp and and a thick mesocarp, to further shorten the unproductive growing period, to slow down and limit the height growth and to increase the fruit content.

The fruitpulp contains 56 % oil on average. As the ratio fruitpulp to nuts depends mainly on climate and variety, the oil content of the fruit (as a whole) varies widely in the range of:

- 14 % for dura fruit in a dry region,
- 20 to 27 % for wild or semi-wild dura fruit, and
- 36 % for tenera fruit.

Processing of the fruit is done as soon as possible after the harvest, because enzymes in the fruit react within 24 hours to form free fatty acids from the oil which substantially reduce the commercial value. The crude (red) palm oil contains a high level of ß-carotene and is an important source of vitamin A in the areas where it grows, preventing the disease xerophtalmia, that can cause blindness.

Palm oil for the world market is usually refined and used for the production of margarine or for other direct cooking purposes. Palm kernels contain 46 % to 48 % oil which is chemically quite similar to coconut oil, the use of which is described further below.

1.2.2 Coconut palm

The coconut palm, *Cocos nucifera,* is botanically grouped in the same subfamily Cocoideae as the oil palm. Although it is long since cultivated in all tropical regions, it originates from the South-West Pacific and South-East Asia; a region which is still the main producer.

For optimal growth, the coconut palm needs an average annual temperature of about 26° C with only small amplitudes between day and night. Therefore, even

Figure 3: Coconut palm
(a) unripe fruit with endosperm beginning to grow, (b) ripe fruit, (c) germ after 3 months, Me = mesocarp, EM = embryo, En = endocarp, Esp = endosperm, Kei = germ, Ha = „apple". Source: S.Rehm, G. Espig, 1984, p. 87

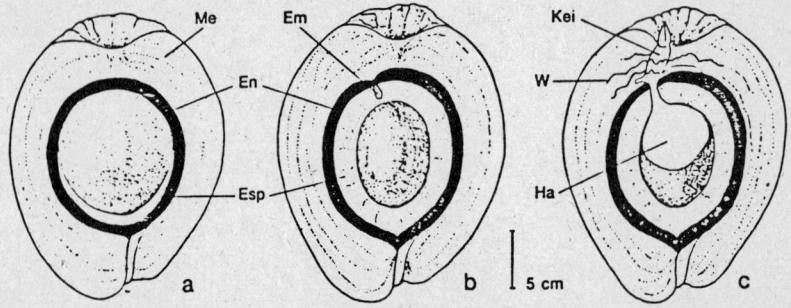

along the equator good yields are only realized below altitudes of 750 m. At sea level, the area of cultivation extends at least 150 north and south of the equator; in the Pacific it even reaches the subtropics. Where the plant depends on rainfall, 1250 to 2500 mm per year are seen as optimal. Good sunshine conditions are also necessary.

World production of coconuts and copra (the dried, but otherwise unprocessed flesh of the nut) has only increased moderately over the last decade and currently stands at about 35 million tons. The „Far East" countries (including India and Sri Lanka) together with the Pacific region produce almost 90 % of this volume, Latin America and Africa share the rest about equally. Most important single producing countries are Indonesia (over 10 million tons p.a.) and the Philippines (about 8 million tons p.a.). For coconut oil, local demand is very high in Indonesia; therefore, the Philippines clearly are the leader in the market with a share of over 50 %. Africa as a whole contributes less than 3 %.

Apart from faster ripening and shorter stems, breeding efforts aim at resistance against diseases (e.g. lethal yellowing which has done tremendous damage in the Caribbean). Crossings with so-called Malayan Dwarfs have given good results in this respect. Major pests are the rhinoceros beetle and a number of other leaf eating beetles and caterpillars.

The cultivation of coconut palms starts with planting the whole fruit, leaving just the upper end above the surface. On germination, the embryo forms a so-called „apple" (which is also consumed fresh). After about 4 to 5 months, the first roots leave the fibrous mesocarp. Planting distances for commercial plantation are about 9 m for high growing and 6 to 7 m for low growing varieties. Undercropping or double-use by grazing is common and can be the most economic land utilization.

A full-grown coconut palm yields 30 to 50 nuts per year with 8000 nuts per ha and year counting as a good harvest. Low growing hybrids usually have smaller nuts but can yield between 200 and 600 fruits per year.

To gain coconut oil, the fibrous husk is often separated from the nut. The nut is then split, usually with a bush knife, the flesh taken out and dried. Drying takes place in the open sun or in simple copra kilns which are fired with the coconut shells. The result is copra which has an oil content of 65 to 70 %. Maximum yields for new varieties are 9 tons of copra per ha, from which 6 tons of oil can be extracted. The actual extraction of the oil from copra is described in other chapters.

Coconut oil contains an extremely small percentage of unsaturated fatty acids. It therefore has a high melting point (22 to 26° C) and does not become rancid. It is therefore highly valued in warm climates and in others used for cakes and pastries. Other than for food purposes, the main use is quality soap.

1.2.3 Soyabean

Although not suited for small-scale extraction of the oil, the soyabean has, since 1945, become the most important source of both vegetable oil and protein and is therefore briefly characterized.

The soyabean or soyabean, *Glycine max*, is a member of the *Papilonaceae*, which includes some forty species of frequently twinning shrubs, distributed generally in the Asia and Australasia region. It is con-

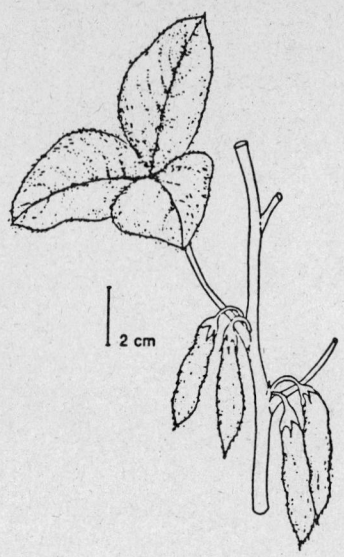

Figure 4: Soyabean.
Source: S. Rehm, G. Espig, 1984, p. 94

As shown in Table 1, the world production of soyabean has more than doubled in the last decade and a half and currently stands at over 100 million tons per year. The USA is the world's largest exporter, and together with China and Brazil, accounts for over 90 % of the world production (Africa: 0.2 %). Highest domestic consumption is in Asia, where it has been a basic food for centuries. Soyabean is mainly cultivated for its seeds commercially used for human consumption, stock food, and the extraction of oil. It is presently the world's most important grain legume in terms of total production and international trade.

sidered as having its origin in north-eastern China, although the genus has two major centres. One is in eastern Africa, the second in the Australasian region with a secondary centre in China. From China soyabean spread to the neighbouring countries Korea, Japan and South-East Asia and finally around the world. As a cultivated crop it remained basically confined to Asia until the beginning of the century, when the USA developed soyabean into a major commercial crop.

The wet subtropics provide the best climate for the soyabean with average annual temperatures of around 25° C and optimal rainfall of 500 to 750 mm per year. The plant is extremely photoperiodic, with most varieties only flowering with day-light less than 14 hours a day. Day- light periods shorter than 12 hours lead to dwarf growth and reduced yields. All varieties are adapted to specific conditions.Cultivation of certain varieties is limited to particular geographic latitudes.

The fruit is normally a short hairy pod, which can vary from 2 to 10 cm in length and 2 to 4 cm in width according to variety. The number of pods per plant can vary from a few dozen to several hundred, depending mainly on climatic conditions. They usually contain three, occasionally more, small, hard or ovoid seeds, usually between 5 and 10 mm in diameter. Seed weight varies considerably in the range of 5 to 40 g per one hundred seeds. The oil content of the seeds, as the major characteristic for the purpose of this booklet, varies between 15 and 22 %, which is, at any event not sufficient for efficient extraction with small-scale technology.

Harvesting can be done by all methods from entirely manual to fully mechanized, depending on cost factors in a concrete context. In regions where labour is plentiful, plants are pulled by hand, thrown into heaps and threshed with sticks. In general, wind-rowing is not to be recommended, except in those circumstances where field conditions preclude natural drying, for instance in some West African areas.

Yields obtained range from 0.5 tons per ha in West and East Africa, 1.0 tons/ha in

Central and southern Africa, 1.5 to 2 tons/ha in East Europe and most of Asia to an average in the USA of 2.5 tons/ha. These relatively low average yields should be judged against yields of nearly 6 tons/ha achieved by commercial growers in the USA, which also indicate the huge potential for increased production in tropical countries.

As mentioned earlier, the cultivation of soyabeans would never be economic without the double potential for vegetable oil and the protein-containing meal which accounts for 40% of the production value. Due to the low oil content, the modern process of solvent extraction is usually applied which, in turn, is only relevant for large-scale industrial operations.

Soya oil normally contains 10% linolenic, 55% oleic and 30% linoleic acid with up to 50% variation in a specific component. Without going into details, one might say that these components make the oil without further processing rather poor and unstable in flavour for direct human consumption. As an industrial rawmaterial, it is mainly used for the production of margarine.

Soyabeans contain a toxic factor which blocks the activity of the digestive enzyme trypsin. Before feeding whole seeds to pigs or poultry, this trypsin inhibitor should be destroyed by heating. Since soyabeans are normally heat treated during processing, oil cake is generally inhibitorfree.

1.2.4 Groundnut

The groundnut, *Arachis hypogaea*, also known as the peanut or earthnut, is botanically a member of the *Papilionaceae*, largest and most important member of the *Leguminosae*.

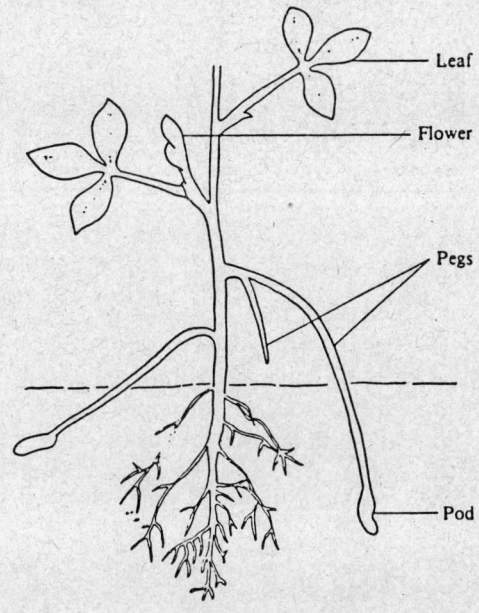

Figure 5: Groundnut. Source: E. A. Weiss, 1983, p. 103

Mainly native to warmer climates, groundnuts frequently provide food for humans or livestock, and, in the absence of meat, form a valuable dietary protein component. The groundnut originates from South America (most likely Bolivia), where a large number of wild species are known to exist. The oldest indications of groundnut cultivation are from the pre-Columbian native societies of Peru. By the time of Columbus, the crop was widely distributed in South and Central America and in the Caribbean. It was probably brought to West Africa from Brazil in the 16th century, from there to the African east coast and so to India. In Africa, groundnuts have become so deeply integrated into society, that traditional customs have arisen around the crop.

Groundnuts grow best at an average temperature of 27°C with 30°C being optimal for germination. Under sub-optimal temperatures, the vegetation period is lengthened by 1 to 2 months. The demand for sunlight is relatively low; a reason why in Africa it is often cultivated in mixed cropping systems together with maize and oil palms. Average annual rainfalls of 500 mm are sufficient for the cultivation. For early varieties, even 200 to 300 mm during the vegetation period are accepted (as in the Sahel region).

Africa normally produces between 25 % and 30 % of world groundnut production and roughly one third of world exports, with Nigeria, Senegal, Zaire and the Sudan (in that order for 1985) being the main producers. Nigeria's leading role in Africa, however, has been put under pressure due to a general decline in agricultural production as a result of the crude oil boom. For groundnut oil, the country has even become the largest African net importer (13 000 tons in 1985). The most important exporting countries for groundnut oil are now Brazil, China and Senegal. The United States are basically a residual supplier to foreign markets, of which the European Economic Community is the most important.

The groundnut is an annual legume, and there is a wide variation in the types cultivated in particular localities. In general, there are two main types which are distinct in appearance: One is upright with an erect central stem and vertical branches, the other has numerous creeping laterals. The first is more commonly grown for mechanized production, the second under peasant farming systems.

Average yields of groundnuts in shells are about one ton per ha with Africa having the lowest results of 0.75 tons/ha and North America ranging over 3 tons/ha. The shells make up 30 % of the weight; the kernels, commonly known as peanuts, can contain up to 50 % oil (although the usual range is 40 % to 45 %) and 25 % to 30 % protein.

Groundnuts give a pleasant tasting oil for direct human consumption and is used as a salad oil or for cooking. The oil is also further processed to margarine or Vanaspati in India.

Improper handling after the harvest can cause the development of poisonous mycotoxins. Groundnuts are particularly susceptable to the development of aflatoxin. Although aflatoxin is insoluble in vegetable oil and is normally concentrated in the cake, impurities accompanying the oil might contain it. Groundnut processors should therefore be especially aware of the danger of aflatoxin.

1.2.5 Sunflower

The sunflower, *Helianthus annuus L.*, is a member of the *Compositae*, a large and successful family of flowering plants occurring throughout the world. The genus *Helianthus* is named from the Greek *helios* meaning sun, and *anthos* flower.

Basically a temperate-zone plant, the main commercial production of sunflower is in the warm-temperate regions but breeding and selection have produced varieties adapted to a wide range of environments. Optimal conditions are short, hot (around 27°C) summers with not too much rain (around 250 mm) during flowering and fructification. Greatest production is between latitudes of 200 to 500 north and 200 to 400 in the southern hemisphere, usually below altitudes of 1500 m.

Sunflower is the only crop of worldwide commercial importance that originates from the area of the USA (probably south-western states to Mexico). Earlier used for food by the Indians of that

Figure 6: Sunflower.
Source: KIT, 1979

region, it was already cultivated when the Europeans arrived in North America. The post-war introduction of Russian varieties, which not only pushed the oil content of the seeds to 50% but were suitable for mechanized harvesting (90 to 150 cm high), had an immense impact on the development of sunflower as a commercial crop in Europe and the Americas.

Today, world production of sunflower ranks fifth among oil seeds and third among the vegetable oils (see Tables 1 and 2). In world trade, it has established an equally important position, although a significant portion of the production in developing countries is for personal consumption and therefore not registered. World production of sunflower seeds was dominated by the USSR which used to account for 50%, but has declined since the 1970's to roughly half that share or 5 million tons in absolute terms. Other major producers are now Argentina (3.4 million tons), China (1.9), France (1.5) and the USA (1.4). Argentina has also become the leading exporter of sunflower oil and now accounts for almost half the world trade.

The cultivated sunflower is a tall, erect, unbranched, coarse annual, with a distinctive large, golden head. The plant grows rapidly, the stem varying in height from 1 to 3m when full grown with individual plants of giant varieties reaching 5m. New hybrids have a shorter stem and are remarkably even in growth and final height (within 5% deviation). The disc-shaped head is born terminally on the main stem and branches where these occur and is commonly 10 to 30 cm in diameter.

The fruit, or sunflower seed, ranges in colour from black through to white, but brown, striped or mottled seed can also occur. Seed varies greatly in size and weight, but is generally a compressed, flattish oblong. Average seed yields are currently 1.3 tons per ha, but can reach up to 4 tons/ha. The oil content of sunflower seed is between 25 and 48%, but can reach 65% under experimental conditions. Important for developing countries is the fact that high temperatures during seed development can reduce the total oil content to below 25%, which would make small-scale processing less rewarding.

Sunflower oil, which is pressed in a cold stage, is a very highly valued salad oil; lower qualities are also used for technical purposes (paints).

1.2.6 Sesame

Sesame, *Sesamum indicum L*, member of the family *Pedaliaceae,* is probably the most ancient oilseed used by man and

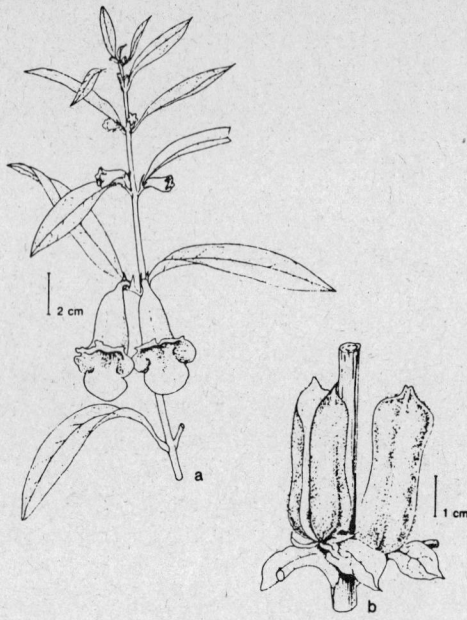

Figure 7: Sesame. (a) shoot top with flowers, (b) ripe capsules. Source: S.Rehm, G. Espig, 1984, p. 102

originates from the Ethiopian area. It occurs as numerous species and is locally known under a variety of names, such as gingelly and til in India, sim-sim in Arab countries and East Africa and benniseed in Nigeria.

Sesame is considered a crop of the tropics and subtropics and normally requires fairly hot conditions, with temperatures around 26°C encouraging rapid germination, initial growth and flower formation. In altitudes below 1250 m, sesame's main distribution is between 250 north and south of the equator, but it can be found further north in China, Russia and the USA and further south in Australia and South America. Optimal rainfall is 500 to 650 mm per year, but since the crop is reasonably drought resistant it can also be planted in relatively arid zones with annual rainfalls as low as 300 mm.

World production of sesame seed has been almost static for 20 years, at 2.4 million tons per year (see Table 1) and almost exclusively originates in developing countries. Major producers are China, India, Burma and Tanzania (in that order for 1985). A large proportion of the sesame seed harvested is, however, neither marketed nor exported but consumed by local producers and therefore often does not appear in statistics for home production. This is particularly true in Africa, where sesame is grown from north to south, but often in such small plots that it is impossible to calculate the total, large though it may be. It is estimated that only 10 % of the total production enters world trade in sesame seed, the largest exporters being China, the Sudan and Mexico, the largest importers Japan, USA and Hong Kong.

Although reaching as high as 2 tons per ha (in Yugoslavia), average yields are only 350 kg of seeds per ha, because sesame is mostly cultivated in arid regions with poor soils. The average seed composition is 45 to 50 % (highly valued) oil and between 19 to 25 % protein. Sesame seed is relatively sensitive to mechanical damage, and even minor damage at threshing can result in an immediate loss of the viability of the oil extraction process.

1.2.7 Rape and mustardseed

Rapeseed and mustardseed are both obtained from species of *Brassica* in the family of the *Cruciferae* which includes some 160 species, mainly annual and biannual herbs. Of rapeseed, the two most important oilseed producers are *B. campestris L.,* which has a fairly wide world distribution, and *B. napus L.,* which is basically restricted to Europe and North Africa. Of mustardseed, *B. juncea* is the most common and known as Chinese or

Indian mustard or rai. Because of the similarity of the species, the present chapter will refer to them as „rape" and summarize the characteristics.

The origin of rape is most likely the South and East Asian region, since the oldest known references to its cultivation are from India, China and Japan. Secondary centres could be in the Mediterranean area. Whereas in the West and East, rape was originally cultivated for its roots and leaves (as a food), in India the seed was selected for its oil, and this started the wider distribution of the crop. Rapeseed's major use then became the production of oil for industry or domestic lighting. As an edible oil, rapeseed was initially only used by poor people, but the development of new technologies has increased its attractiveness for human consumption and animal feed. Mustardseed has long been used for spices.

Today, rapeseed and mustardseed rank with about 19 million tons sixth in world production of major oil plants and with 6 million tons fourth in vegetable oils (see Tables 1 and 2). Major seed producers are China, Canada, India and France (in that order for 1985), main oil producers are Europe, Canada and Japan. World trade in rape and mustard oils has been steadily increasing since the 1960's with volumes currently around 1.3 million tons. Surprisingly, with 75 % of the trade, exports are dominated by European countries (Federal Republic of Germany, France and others); other major producers, like Canada, are still increasing their market share. Production and exports from Africa are negligible, North African countries, in fact, importing one third of all commercially available quantities.

Oilseed rape and mustard are basically temperate crops which prefer moderate

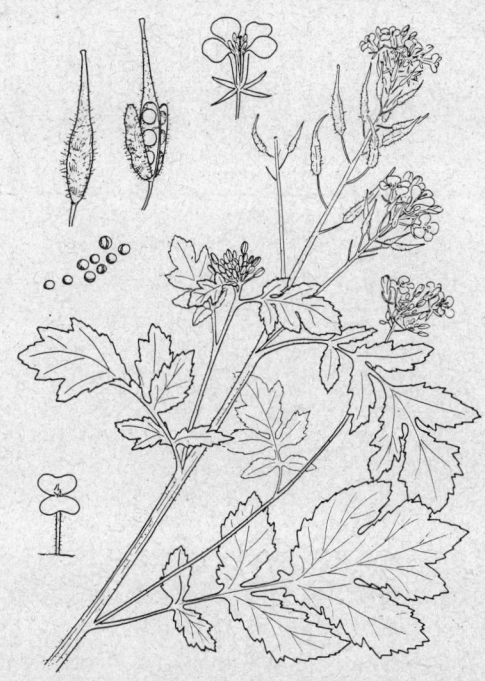

Figure 8: Mustard.
Source: KIT, 1979

temperatures below 25°C during growth. Breeding and selection has considerably increased the geographic range of cultivation with hardy varieties being able to withstand long periods of snow cover and very low temperatures and others able to withstand more than 40°C for a limited period during the vegetative phase. Optimal rainfall is considered to be 700 mm per year. Rape will still produce a good crop using mainly residual soil moisture, provided some rain falls between planting and the seedling stage and at main flowering. These characteristics are of particular advantage in tropical regions of high altitude with significant temperature variations and low rainfall.

Yields per ha can reach more than 3 tons of seeds under optimal conditions (Europe), but on average yields are just over

1.2 tons/ha. The protein content of the seeds varies from 10 % to 45 %, the oil content is normally in the range of 30 % to 50 %, but can reach up to 60 %.

Rapeseed oil is dark, but after refining becomes light yellow and resembles sunflower oil. The colour is influenced by the seed chlorophyll level. A low value produces a light coloured oil which is commercially desirable. In the past, oil produced from the higher yielding varieties contained high levels of erucic acid, which constitutes a health risk for human consumption. Breeding has led to varieties without this acid. Oil from older varieties is mainly used for technical purposes.

1.2.8 Other oil-yielding plants

The **sheanut** or **karité**, *Vitellaria paradoxa* (syn.: *Butyrospermum parkii*), is a wild-growing tree in the countries south of the Sahel zone. The tree is relatively small, growing to a height of some 12 metres, having a thick trunk and a number of spreading branches which form a dense crown. It bears fruit after 10 to 15 years, reaching full bearing capacity at 20 to 25 years.

From Senegal to Sudan, the fruit, which requires 4 to 6 months to mature, is locally of substantial economic importance in particular for rural women who collect the nuts and process them to a butter-like edible substance. Average yield per tree is 15 to 20 kg of fresh fruit, with one third of the trees being productive each year. 50 kg of fresh nuts give 20 kg dry kernels which yield about 4 kg of shea butter.

The kernel of the nut contains 32 to 54 % fat. In international trade, shea butter serves as a cheaper substitute for cocoa butter and is in high demand when prices for the latter are up. It is also valuable for the cosmetic industry. With an estimated total production (collected nuts) of 500 000 tons per year, Nigeria, Ghana, Ivory Coast and to a lesser extent, other countries in West Africa, are reported to export altogether about 30 000 tons.

Actual production and even export figures, however, are difficult to verify. In Mali, for example, production of kernels in the season 1985/86 is estimated to be 44 000 tons, of which 10 200 tons were officially exported. Inofficially, an estimated additional 16 000 tons have left the country for unknown destinations.

Cotton, *Gossypium spp.*, is basically a textile plant, which is mainly cultivated for its hair or lint (9.5 to 20 mm long) on the stem and leaves. Apart from the commercial value of the fibre, however, the seeds contain about 25 % oil thus making cotton the third most important oil crop. Of additional value is the protein-rich (40 %) oilseed cake, which gives a good protein supplement for cattle. The oilcake, however, can contain free gossypol, a pigment sometimes found in cottonseed. In this case, the use of the cake is limited to feeding to non-ruminants in small quantities.

World production of the seeds has exceeded 30 million tons in recent years, and of cottonseed oil between 3 and 4 million tons (see Tables 1 and 2). Major producers of cottonseeds are China, USSR and the USA. Although the plant originates from semi-arid zones in Africa, the continent as a whole (mainly Egypt and Sudan) only contributes 7 % to world production. World trade in cottonseed oil is dominated by the USA and Brazil. Due to the relatively low oil content of the seeds, solvent extraction is generally applied as a larger industrial operation; small-scale equipment does not appear to provide economic extraction.

Linseed, *Linum usitatissimum L.*, is also mainly cultivated for its fibre, but is in multiple use as a spice and an oil yielding plant. World production in linseeds has considerably decreased over the last two decades and currently stands at about 2.5 million tons annually. The major producers are Canada and Argentina; production in Africa is negligible. Linseed oil is used mainly for technical purposes (paints) and faces heavy competition from other vegetable oils and in particular synthetic products. The market for the oil is more or less taken by Argentina (71 %) and the Federal Republic of Germany (17 %). Small-scale extraction is done in India.

Maize (Corn in American diction), *Zea mays L.*, is basically a starch plant, but the oil content of the germ is also used on a commercial scale. The plant originates from Central and South America from where it began to spread to almost all parts of the world. Today, the USA is the largest producer, both American sub-continents together still contributing the majority (60 %) to the total world production. Maize oil is rather a minor by-product of the maize starch manufacturing industry and is extracted on an industrial scale. World trade in maize oil amounts to about 300 000 tons annually, largest exporters being Europe and the USA. Africa accounts for net imports of some 10 000 tons a year.

Safflower, *Carthamus tinctorius*, originates from the eastern Mediterranean and the Persian Gulf and was first cultivated for the orange dye obtained from the florets. The seeds of modern varieties, however, contain 36 % to 48 % oil with good qualities for human consumption; a fact which now constitutes the main purpose for the cultivation of the plant whose seeds are suitable for small-scale processing.

The plant is a highly branched, herbeceous annual, varying in height from 30 to 150 cm and generally has yellow flowers. The fruit resembles a small, slightly rectangular sunflower seed, but with a thicker, more fibrous hull. It gives best yields (but with 1 ton/ha still low) in semiarid climates. Due to its fairly good drought and salt resistance, safflower is suitable for areas where other oil seeds are difficult to grow.

World production of safflower seed increased sharply in the 1960's but has since remained static at just below 1 million tons a year, with India accounting for more than half of the total. Safflower is still considered a typical small-holder crop and is mostly processed and consumed locally; for large plantations, as those in Mexico until recently, other oil seeds, particularly hybrid sunflower, appear to be more profitable.

Castor, *Ricinus communis*, is indigenous to eastern Africa, most likely originating in Ethiopia, but has a contemporary world-wide distribution in the warmer regions. Originally a tree with heights up to 10 m, most mechanically cultivated varieties today are short-lived dwarf annuals and only 60 to 120 cm high. Castor plants also grow wild in many countries and production could be substantially increased if incentives were sufficient. This is particularly so in eastern Africa, from the southern Sudan to southern Tanzania, where there are extensive areas with wild species.

Castor seeds (or „beans") almost exclusively originate from the developing countries, the major producers being China and Argentina. Although Africa does not play a major role in official statistics of world production (3 %), castor seeds might well be of greater economic significance in the local context. In world trade of castor oil, Brazil and India clearly dominate the supply side with a combined market share of over 90 %; largest importing region is Western Europe.

A large proportion of castor seeds on local markets in less developed regions is obtained from wild or semi-cultivated plants. At least in Africa, systematic cultivation of pure stands of castor by peasant farmers is the exception. More often castor is interplanted with other crops, sown along borders of fields, planted in areas unsuitable for other crops or merely protected when found growing naturally. In a cultivated stage, yields per ha are a mere 0.5 to 1 ton. The seeds, however, have a high oil content of 42 % to 56 % and are suited for small-scale processing methods. The oil is used for a variety of technical purposes.

Other oil yielding plants suitable for small-scale processing, which, for reasons of space cannot be dealt with in detail, include:

– **Physic Nut** (Purgier), *Jatropha curcas*, the oil of which is mostly used for soap and might be economically used as a fuel on the Cape Verde islands,

– **Niger Seed**, *Guizotia abyssinica*, which is produced in India and Ethiopia and gives a good edible oil,

– **Babassu**, *Orbignya oliefera*, originating from Brazil with nuts containing 2 to 8 kernels with 60 % oil similar to coconut oil,

– **Cohune**, *Orbignya cohune*, growing in Central America, the nuts containing a kernel with 60 % oil comparable to coconut oil,

– **Neem**, *Melia azadirachta L. (Azadirachta indica)*, growing in Africa, SE-Asia and India, with seeds containing 45 % oil, which is mainly used for soap and medical purposes,

– a large number of **wild growing oil yielding plants** of local importance.

1.3 By-products

In the above characterization of the major oil plants, reference has been made to the

Table 6: **Important Oilcrops and their By-Products**

Oilcrop or (intermediate) product	By-Product	Use
Oilpalm, fruit bunch Oilpalm fruit Palm oil and	bunch fibrous residue and sludge	fuel fuel/fertilizer roughage traditionally for human consumption animal feed
Palmnuts Palmkernels Palmkernel oil	shells palmkernel cake	fuel/charcoal animal feed

Table 6 continued

Oilcrop or (intermediate) product	By-Product	Use
Coconut Husked coconut Shelled coconut Coconut oil	husks shells coconut fibre (traditionally) or coconut cake	coir for matting fuel/charcoal animal feed
Soyabean Soyabean oil	soyabean cake	human consumption or animal feed (when trypsininhibitor free)
Groundnut Shelled groundnut Groundnut oil	shells groundnut cake	mulch/litter particle board human consumption or animal feed
Sunflower Sunflower kernels Sunflower oil	husks sunflower cake	fuel/filling material, polishing material, roughage animal feed
Sesame Sesame oil	sesame cake	human food or animal feed
Rape/Mustard Rape/Mustard oil	cake	animal feed
Castor bean Castor oil	cake	fertilizer if detoxified as animal feed
Cotton seed Cotton seed oil	cake	animal feed (limited by free gossypol)

main use of the crops, i.e. in most cases the extraction of vegetable oil which, in turn, is used mainly for food and, to a lesser extent, for technical purposes.

In the present chapter, the by-products obtained from processing the above oilplants for oil are summarized (see Table 6).

High protein containing oilcakes are much too rich to be fed directly to animals. They have to be mixed with starch and fibre containing feedstuffs in order to be properly digested. Handbooks on animal husbandry will provide detailed information on how a proper animal feed can be prepared with oilcakes as ingredients.

In Table 7, some examples are given of the composition of locally produced oilcakes of important oil seeds.[1]

[1] For more information see, for example: Göhl, B., Tropical Feeds, FAO Animal Production and Health Series No. 10, FAO, Rome 1981.

1.4 Further processing

Vegetable oils and fats can be consumed directly or be the rawmaterial for further processing. Below, some of the more important products derived from vegetable oils are discussed. Most of these products are manufactured by a specialized industry.

Refined salad and cooking oils

In food, vegetable oils are normally preferred with a bright colour and a bland taste. Neutrally tasting salad and cooking oils are generally made from peanut, sunflower, sesame, maize or cotton seed oil by a complex refining operation which includes neutralization, bleaching and deodorization. Soyabean oil and rape seed oil are less suitable as they possess a characteristic flavour after deodorization.

Margarines

Margarines are an emulsion of fats in water, in which the water compound is dispersed as in butter made from milk. The fat needs to have a certain plasticity

Table 7: **Examples of the Chemical Composition of Oil Cakes fit for Animal Feed**

Oilseed cake	Protein	Fibre	Carbohydrates	Fat	Ash
Palmkernel	16	29	28	23	4
Coconut	20	9	48	18	5
Soyabean	44	8	33	8	7
Groundnut					
with hulls	34	26	25	10	5
dehulled	49	5	32	9	5
Sunflower					
dehulled	34	13	32	14	7
Sesame	35	8	28	17	12
Rape	36	10	32	13	9
Cotton					
with hulls	30	8	47	8	7

Source: KIT

and to melt readily in the mouth. Such a fat is obtained by mixing, for instance, coconut and/or palmkernel oil with other oils which are hydrogenated to an appropriate degree.

As the product contains water it is easily spoiled by bacterial contamination or oxidation. Margarines need therefore to be tinned or stored under refrigeration.

Pure oils and fats can be kept much better. In India and other tropical countries butter is therefore processed into a pure oil, known as ghee.

Bakery products

Fat is a highly necessary ingredient in baked goods as it not only contributes to the flavour but also to the physical structure of the product. To provide the fats with the required so-called „shortening" value, a large range of special products are on the market for the bakery industry which are made by the hydrogenation of vegetable oils and compounding of fats from different origins.

Soaps

Soaps are commonly made by boiling a fat with strong lye. The effectiveness of such a soap depends mainly on its surface activity properties and its solubility. Both depend largely upon the length of the chain and the unsaturation of the fatty acids which form the soap.

Soaps made from the higher fatty acids (e.g. stearic) are very efficient detergents, but have only limited solubility, which limits their usefulness as a household soap. Soaps of the lower fatty acids (e.g. lauric acid) are freely soluble but are less efficient as a detergent.

An optimum balance can be reached on the basis of a fat from animal origin or palm oil, mixed with 15 % to 30 % of coconut or palmkernel oil.

Lubricants

Mineral oils have superseded „fatty" oils (i.e. animal and vegetable oils) as lubricants as they are more stable and cheaper. However, fatty oils have certain special advantages to mineral oils, since they have a superior ability to cling to metal surfaces in a thin film. Suitable fatty oils are those which are sufficiently saturated to be stable. They are used for lubricating light machinery. Castor oil is more viscous than ordinary oils and hence is suitable for lubricating heavy machinery. The latter is also used as a base for fluids in hydraulic systems.

Illuminants

Cheap illuminants from petroleum have virtually eliminated vegetable oils as burning oils, except in isolated regions. However, candles based on paraffin wax or beeswax need a hardening agent to assist in maintaining their shape in hot weather or to burn without dripping. Next to stearic acid, very highly hydrogenated oils can be used for this purpose.

Other

Stable fats and oils can be used in cosmetics such as, for instance, palm oil or shea nut butter in creams and castor oil in alcohol-based hair dressings. In pharmaceuticals they can be used to carry fat-soluble substances as vitamin concentrates. They can also be used to prevent infestation of crops by insects.

Highly unsaturated oils are used as drying oils in paints and varnishes. Linseed oil is important in this regard, followed by soyabean oil. Castor oil can be chemically dehydrated to give a fast drying oil, suitable for use in water-resistant varnishes and enamels.

2. Target Groups and Technologies

2.1 Family level

In most societies, the family comprises the smallest economic unit. In developing countries, however, a family does not usually consist simply of husband, wife and their children. More often, a family includes other relatives (grandparents, grandchildren, cousins, etc.) or even close friends living in the same household. In islamic countries, a family might include a second or more wives and their children. In traditionally oriented societies, the extended family is still very important for the social identity of the individual.

Viewed from the point of oil processing, a family is defined here as a group of people living together in one household, ranging in numbers from one (exceptionally) up to 30, sharing common social and economic interests and usually having their meals together. Oil processing at this level of social aggregation primarily aims at subsistence needs, but also contributes to cash income.

In many areas of developing countries, oil fruits and oil seeds are available as a rawmaterial, and processes to prepare vegetable oil are known to the population. The modest needs for vegetable oil for the family pot are supplied by the women, following traditional methods of processing.

Next to the production of oil for the family, women also make products for sale on the local market to earn the money required to pay for their contributions to other family needs, such as the ingredients for the daily family pot. These are generally made up by some vegetables and spices and, when the money is available, by some (dried) fish or meat, all prepared into a sauce. This sauce is eaten as a relish, that accompanies the staple, which is based on a starchy food as a grain or a root crop.

When men produce an oilcrop that must be processed before marketing, the processing is generally not carried out at the family level. When sizable quantities are involved, the processing is carried out by specialized groups at the village level. This will be dealt with in the next chapter.

Below, a description is given of some traditional methods at family level to process oil palm fruit and oil seeds for food and cash income.

2.1.1 Oil palm fruit
(see Flowsheet 1)

The semi-wild palms are mainly of the „dura" variety. The dura fruit contains large nuts with a thick outer shell. Around the nut is a relatively thin layer of oil-containing fruit pulp.

Women search for fruits that have dropped out of the bunches from the trees or buy loose fruit at the market. Women do not climb palms and thus can only obtain bunches of fruit that have been cut by the men. Women can pay men to climb the

Flowsheet 1 Traditional Process for Oil Palm Fruit

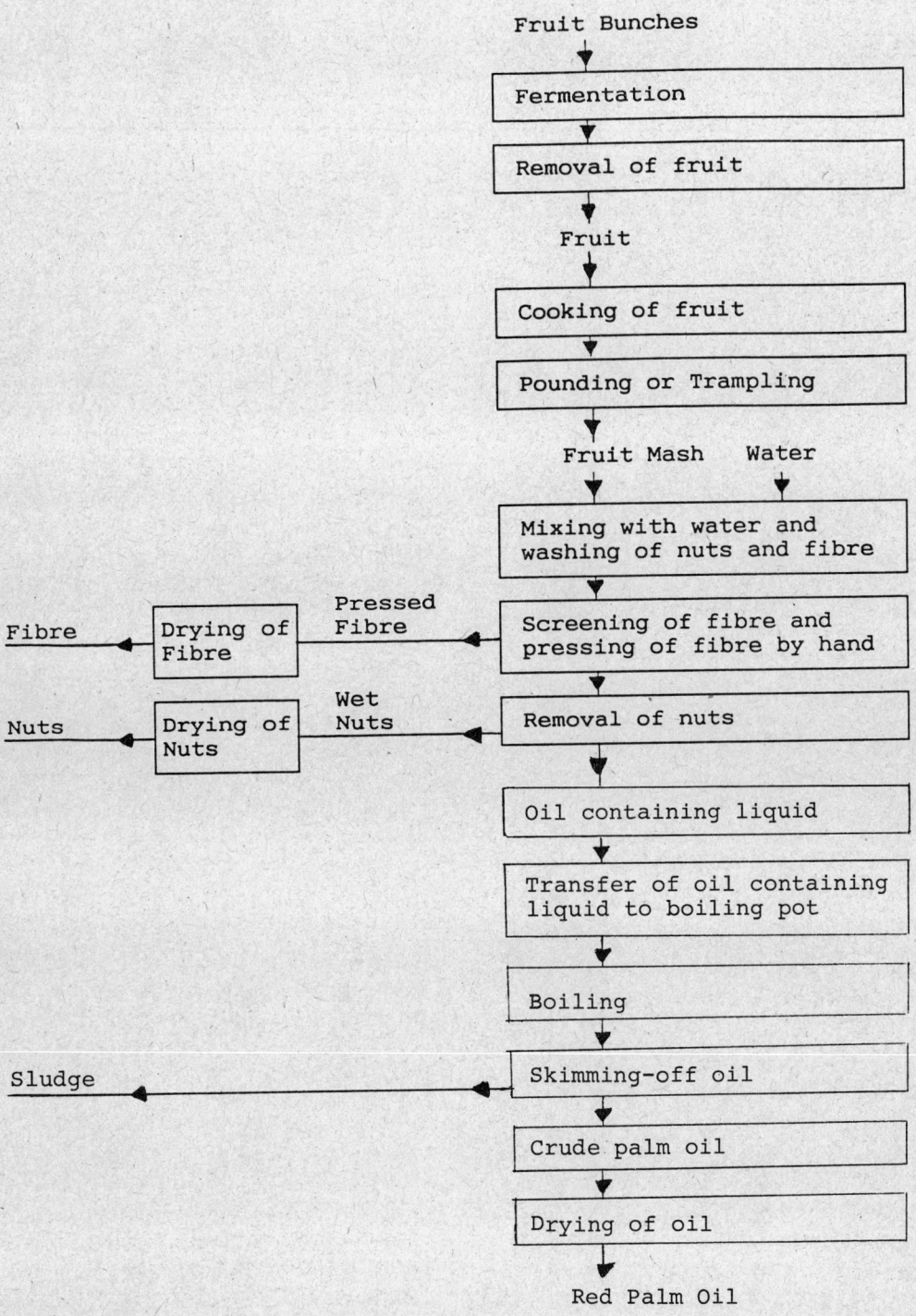

trees and cut the bunches. Bunches have to be stripped to obtain the fruit.

The fruit can be more easily separated from the bunches after fermentation in a heap for 3–4 days. To facilitate the separation the bunches can be cut in small clusters before fermentation.

The fruit is cooked and subsequently pounded using mortar and pestle. When larger quantities are to be processed, in some areas, the fruit is pounded using a halved drum and a large number of pestles or the fruit is mashed by trampling with the feet in a pit. Men can assist with the stripping of the bunches and the mashing of the fruit. After crushing by pounding or trampling, the mass of fruit pulp and nuts is mixed with excess water. The nuts are washed free from pulp and are allowed to settle to the bottom. The fibres are then thoroughly washed with water and finally pressed out by hand to remove all oil and oil-containing cell material.

In some areas just the floating cream is collected, whereas in others all the liquid that remains after removal of the nuts is taken. This mass is transferred to a drum and boiled for a few hours. The palm oil at the top is skimmed off and finally purified and dried by heating in a separate pot. The remaining sludge is sometimes concentrated by boiling and used for food.

The nuts are spread on the ground and dried in the sun, after which they are cracked to obtain the kernel, traditionally by tapping between stones. The fibre is also dried and used as a combustible.

The required time reported for processing one drum (44 gallons or 200 l), containing around 150 kg of palmfruit, is on the average 24 and 32 man working hours respectively for Benin and Gambia.

Oil recovery out of a drum varies between 9 kg for dura oil palm fruit in Gambia, about 15 kg for dura fruit in other countries and as much as 20 kg for a dura/tenera (improved variety) mixture in Cameroon.

Traditionally, the fruit could be left to ferment for days, making the processing quite easy. However, this oil contains a high percentage of free fatty acids and has a sharp taste. It is therefore known as „hard" oil.

Possibilities for improvement

In principle the traditional method for processing oil palm fruit is based on the separation of the oil-containing cell material from the nuts and the fibre, followed by the extraction of the oil from the cells by prolonged cooking. This process has a limited oil recovery and requires much water and energy.

These drawbacks can only be overcome by changing to a process that uses practically no water but demands thorough preparation of the fruit, before the oil can be extracted with a hand press.

Since such a system requires much more investment than the traditional process, it is not a feasible proposition at the family level. Only at village level can the investments be justified either in the form of a service system, to be used by processors against a fee, or as an asset of a specialized informal or formal co-operative. Details of the processes are given under 2.2.1.

2.1.2 Oil seeds
(see Flowsheet 2)

Groundnuts

Groundnuts are almost exclusively processed in combination with the utilization of the residue for human consumption. In fact often the by-product, a kind of a

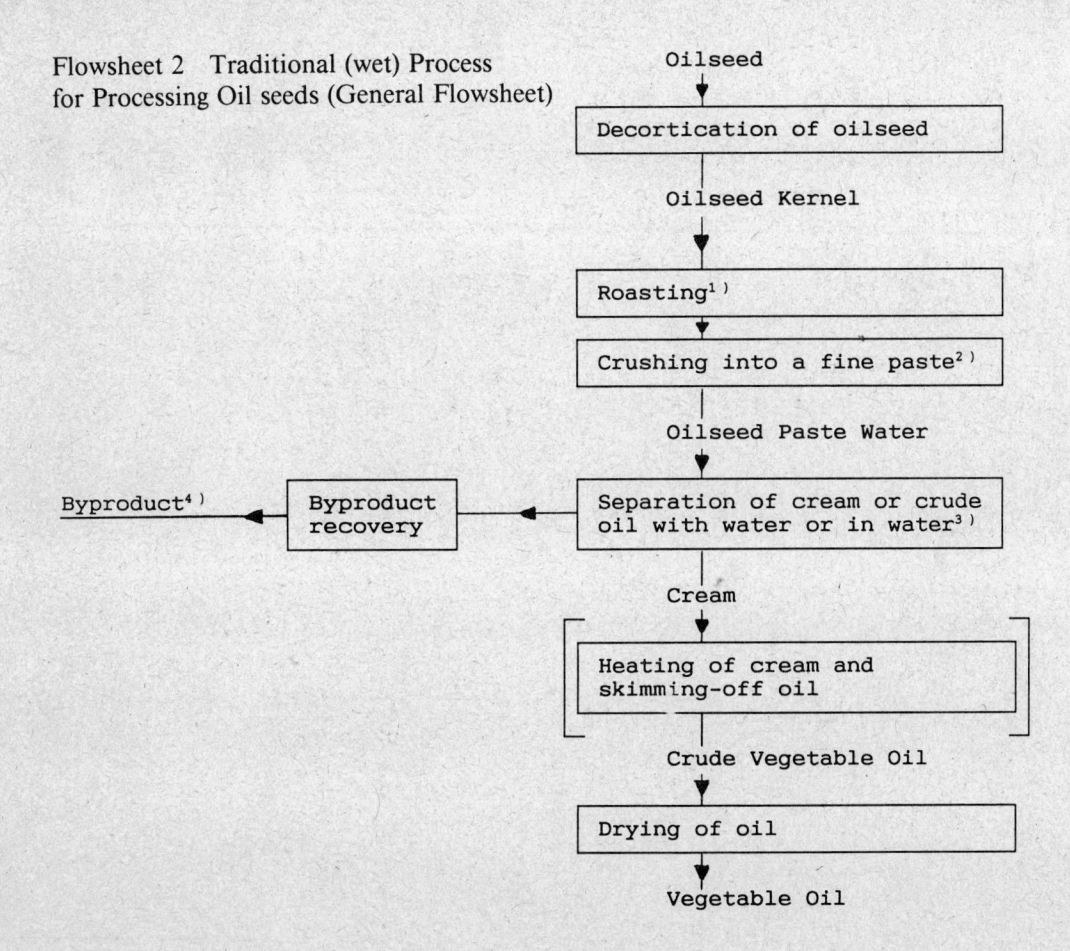

Flowsheet 2 Traditional (wet) Process for Processing Oil seeds (General Flowsheet)

```
                          Oilseed
                             ↓
               ┌─────────────────────────────┐
               │  Decortication of oilseed   │
               └─────────────────────────────┘
                             ↓
                      Oilseed Kernel
                             ↓
               ┌─────────────────────────────┐
               │  Roasting[1]                │
               └─────────────────────────────┘
                             ↓
               ┌─────────────────────────────┐
               │  Crushing into a fine paste[2] │
               └─────────────────────────────┘
                             ↓
                  Oilseed Paste Water
                             ↓
  Byproduct[4]  ┌──────────┐   ┌─────────────────────────────────┐
  ◄─────────── │Byproduct │◄──│ Separation of cream or crude    │
               │ recovery │   │ oil with water or in water[3]   │
               └──────────┘   └─────────────────────────────────┘
                             ↓
                           Cream
                             ↓
               ┌─────────────────────────────┐
               │  Heating of cream and       │
               │  skimming-off oil           │
               └─────────────────────────────┘
                             ↓
                    Crude Vegetable Oil
                             ↓
               ┌─────────────────────────────┐
               │  Drying of oil              │
               └─────────────────────────────┘
                             ↓
                     Vegetable Oil
```

[1] With groundnuts, palm kernels and often with shea nuts.
[2] By pounding, crushing between stones or a stone and an iron bar or by the service mill; fresh coconut is grated.
[3] Groundnut paste is treated by stirring and addition of some water (12 %). Crushed palm kernels are cooked in excess water. Shea nut paste is treated by beating in air and washing of the cream or cooking in excess water.
[4] Tunkusa and kuli-kuli for human consumption from groundnuts; animal feed from palm kernels and coconuts.

snack, has to be understood to be the main product and the manufacturing of the groundnut oil only as part of the process.

In Ghana the following process was observed:
– decorticated groundnuts are roasted, treated by a rubbing action and winnowed to remove the pellicles
– the nuts are crushed between stones, several times to obtain a fine paste
– this paste is stirred vigorously, while gradually adding some hot and/or cold water (about 10 % w/w)
– when the oil appears it is skimmed off and the mass formed into large balls and some more oil is pressed out by hand; the balls are called: tunkusa
– the tunkusa is subsequently processed

into kuli-kuli, a ring or ball-shaped snack, prepared by frying products moulded out of tunkusa in groundnut oil. It can also be used as the main ingredient for „groundnut soup".

For Burkina Faso, a similar process has been described.

A typical example has shown a recovery of 0.5 kg oil and 3.5 kg kuli-kuli (out of 3.8 kg tunkusa) from 4.0 kg of groundnuts; the production of the oil and tunkusa took 5 man working hours.

Palm kernel

After cracking the palmnuts, the palm kernels can be separated out. Traditionally they are processed into an oil after roasting. The roasting makes the palm kernels brittle and more easy to crush by pounding. However, the quality of the oil deteriorates because of the temperature and the oil becomes dark coloured. After roasting the kernels are pounded. Then the pounded mass is mixed into excess water and boiled for hours, during which the oil is skimmed off. Finally the oil is dried by heating. 18 kg of palm kernels give 1 gallon or 4 kg of oil in about 12 man working hours.

Coconut

The basic way to process fresh coconut is to cut the coconut lengthwise in half and to remove the white kernel or so-called 'meat'. The meat is first grated on a grating surface by hand then mixed with water and pressed out by hand or foot. This procedure is repeated several times. The coconut milk obtained is left to stand for a few hours to permit the separation into a supernating oil-containing cream and water. Subsequently the cream is collected and transferred into a cooking pot and heated under continuous stirring to dry the oil by boiling. The protein in the cream coagulates and dries. The oil is filtered. The residue can be eaten as a snack.

Shea nut

Also shea nuts are processed following a wet process. This process has been studied in Mali and includes:
– drying and roasting of the nuts
– decorticating of the nuts
– pounding of the shelled nuts into a liquid mass, that contains particles smaller than about 3 mm
– crushing of the mass into a very fine paste between stones or a stone and a metal bar
– in some areas the brown paste obtained is mixed into water, and air is brought in by beating, a cream appears; this cream is washed several times to remove all brown particles and transferred into a cooking pot; the cream is heated until the oil is collected at the top; the oil is skimmed off
– in other areas or later in the year the brown paste is mixed into boiling water and boiled for an hour, after which the oil is skimmed off; more water is added and boiling continued; finally, a second layer of oil is skimmed off
– by the next day the oil has hardened into a fat, and can be packed in leaves.

The time required for processing (grinding and oil extraction) of 12 kg shea kernels was found to be at least 18 man working hours, of which 14 hours were required for grinding. Generally, however, much more time is needed. Oil recoveries depend on the quality of the nuts and the skills of the women and can range from 20% to more than 40% on kernel weight. Normal yields have been reported to be between 20 and 30%. In South Mali, recovery rates of 34 to 41% have been reported recently.

Possibilities for improvement

Traditional methods make use of readily available utensils as a pounding mortar, crushing stones, calabashes and cooking pots. As they all apply water to assist the separation of the oil, these methods have in common that crushing of the seed into a very fine paste is essential. This stage is the most time consuming and exhausting one. This drawback of traditional methods can only be overcome by crushing using mechanical means. Such means, as motorized mills, require considerable investment and are only feasible at the village level. In fact they are already available in many villages in the form of so-called „service mills".

The only way to avoid the use of a motorized mill is to change to a complete dry process using a hand press. During this process the seed is treated before pressing with care, by crushing into flakes, moistening and heating, in order to make the oil available so that it can easily be extracted by pressing.

The equipment to be used is unsophisticated and sturdy. However, investments required are only feasible at the village level. The process will be treated in detail under 2.2.2.

2.2 Village level

In most rural areas of developing countries, the village is not just a conglomeration of huts and houses but still a functioning community with traditional cultural values and, to a certain extent – common economic interests. Within a village, people might often or regularly come together to do work which is more easily or more effectively done in a sizable group than individually or with family members.

Referring to oil processing at village level, the need either for a specialized trade or for the people to cooperate in groups arises when the quantities to be processed become larger. In West Africa, one comes across groups, processing oilpalm fruit, mainly consisting of women. On the Indian subcontinent, one will find the village oil man, operating his animal-drawn gahni for the processing of oil seeds. These systems are generally operated on a service basis as so-called „service mills", processing the rawmaterials for the client against payment in cash or in kind.

In case the quantities concerned guarantee enough supply, investments in equipment with improved oil recovery or with a labour saving effect can become profitable. The ownership of this improved technology is usually in the hands of individuals, but in the framework of development efforts, self-help groups, pre-cooperatives and more formal cooperatives have been encouraged to establish oil processing units.

However, as mechanized equipment tends to be dominated by men, a shift from processing activities from many women, who are the traditional processors, to a few men can be the result. To make it possible for women to stay in business, the emphasis in recent years in improving traditional technologies has been put on hand-operated equipment.

Below, the existing and possible systems for the processing of oil crops at the village level are presented.

2.2.1 Oil palm fruit
(see Flowsheet 1)

In West Africa, specialized groups can be found when larger quantities of oil palm fruit have to be processed. These groups are well organized and have been reported to be able to process 3-6 drums (450-900 kg) of fruit per day, using traditional methods.

Handpresses

The main drawback of the traditional process with its large water consumption, has been eliminated by the use of presses. These were initially modified wine presses, e.g. the so-called „Duchscher" curb press, which was at one time built in Luxembourg.

Only in Nigeria were these presses made available as privately owned service mills all over the country. The owner of the fruit could come with his fruit and his crew to use the equipment of the mill (as boiling drums, pounding mortar, the press and a clarification drum) or he could bring the fruit to have it processed by the mill. In this last case the owner of the mill would process the fruit with his family. Later on hydraulic presses were introduced but not accepted.

Elsewhere, groups working in the field of Appropriate Technology, took up the design of presses to be manufactured from locally available materials. These groups are for example TCC (Technology Consultancy Centre, Kumasi, Ghana), who designed a press with the spindle in

Figure 9: Duchscher Curb Press

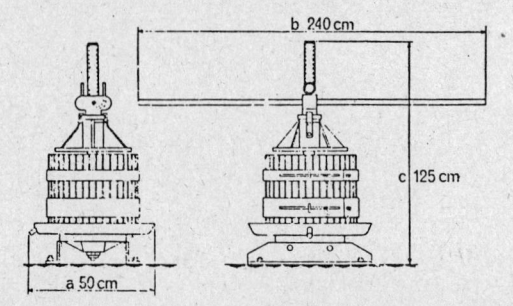

the centre, and ENDA (Environnement et Développement du Tiers Monde, Dakar, Sénégal), who adopted an already existing design for dissemination. Other countries from which activities have been reported are Cameroon, Liberia, Sierra Leone, and Togo.

Hand presses make the traditional process simpler (see Flowsheet 3). However, the need for intensive pounding remains. In order to facilitate the pounding step, KIT (Koninklijk Instituut voor de Tropen, Amsterdam, the Netherlands) has introduced the reheating step, during which the fruit pulp is completely digested. Pounding is only required to remove the pulp from the nuts. The actual digesting of the fruit is carried out by steaming, during which the cell walls are weakened, the protein in the cells denatured and the micro oildroplets, as originally present in the cells, combined to larger droplets, which are more easily pressed out.

A complete process has been designed, starting with the steam sterilization of the bunches to the final drying of the oil, to improve upon its storage properties. When the fruit is cooked in water, the intercellular cement dissolves, giving a suspension of cells (still intact), from which it is difficult to obtain the oil. This problem is avoided by the designed steam sterilization process. To be able to process dura oil palm fruit with a good recovery, fibre has to be recycled and mixed into the mass to be pressed, to prevent the nuts from touching. As the ratio of fruit pulp to nuts in Tenera fruit is higher, it can be processed without recycling the fibre. However, to obtain maximum oil recovery, it is recommended to reheat and press the fibre (or the fibre/nut mixture) a second time. The complete process, including all process steps, is presented as Flowsheet 4. With this process it

Flowsheet 3 Usual Process for Oil Palm Fruit with Hand Press

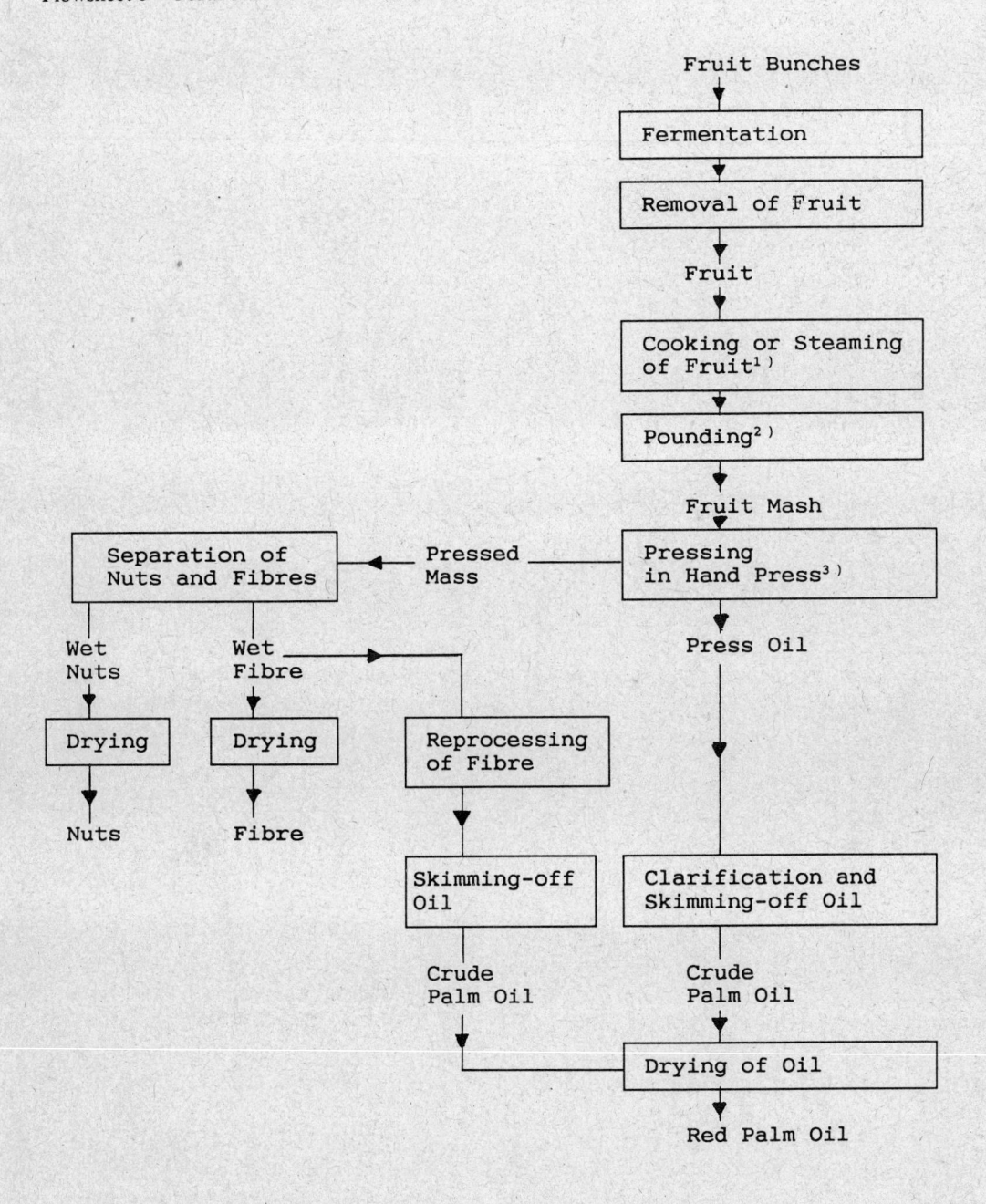

```
                                          Fruit Bunches
                                                │
                                                ▼
                                        ┌─────────────────┐
                                        │  Fermentation   │
                                        └─────────────────┘
                                                │
                                                ▼
                                        ┌─────────────────┐
                                        │ Removal of Fruit│
                                        └─────────────────┘
                                                │
                                              Fruit
                                                ▼
                                        ┌─────────────────┐
                                        │ Cooking or Steaming
                                        │ of Fruit¹⁾      │
                                        └─────────────────┘
                                                │
                                                ▼
                                        ┌─────────────────┐
                                        │  Pounding²⁾     │
                                        └─────────────────┘
                                                │
                                            Fruit Mash
                                                ▼
  ┌──────────────────┐   Pressed      ┌─────────────────┐
  │  Separation of   │◄──Mass─────────│   Pressing      │
  │ Nuts and Fibres  │                │ in Hand Press³⁾ │
  └──────────────────┘                └─────────────────┘
       │         │                            │
     Wet       Wet                         Press Oil
     Nuts     Fibre ──────────┐                │
       ▼         ▼            ▼                │
   ┌──────┐  ┌──────┐  ┌──────────────┐        │
   │Drying│  │Drying│  │ Reprocessing │        │
   └──────┘  └──────┘  │  of Fibre    │        │
       │         │     └──────────────┘        │
       ▼         ▼            │                │
     Nuts     Fibre           ▼                ▼
                       ┌──────────────┐  ┌──────────────────────┐
                       │ Skimming-off │  │ Clarification and    │
                       │     Oil      │  │ Skimming-off Oil     │
                       └──────────────┘  └──────────────────────┘
                              │                   │
                            Crude               Crude
                          Palm Oil            Palm Oil
                              │                   │
                              └─────────┬─────────┘
                                        ▼
                                ┌───────────────┐
                                │ Drying of Oil │
                                └───────────────┘
                                        │
                                        ▼
                                  Red Palm Oil
```

[1] Steaming is advised to prevent the decomposition of intercellular cement.
[2] Pounding can be mechanized for example with the TTC palm fruit pounder.
[3] To be carried out with any press, sturdy enough to press mixtures of nuts and fibre (see 5.1).

Flowsheet 4   KIT Process for Oil Palm Fruit with Hand Press

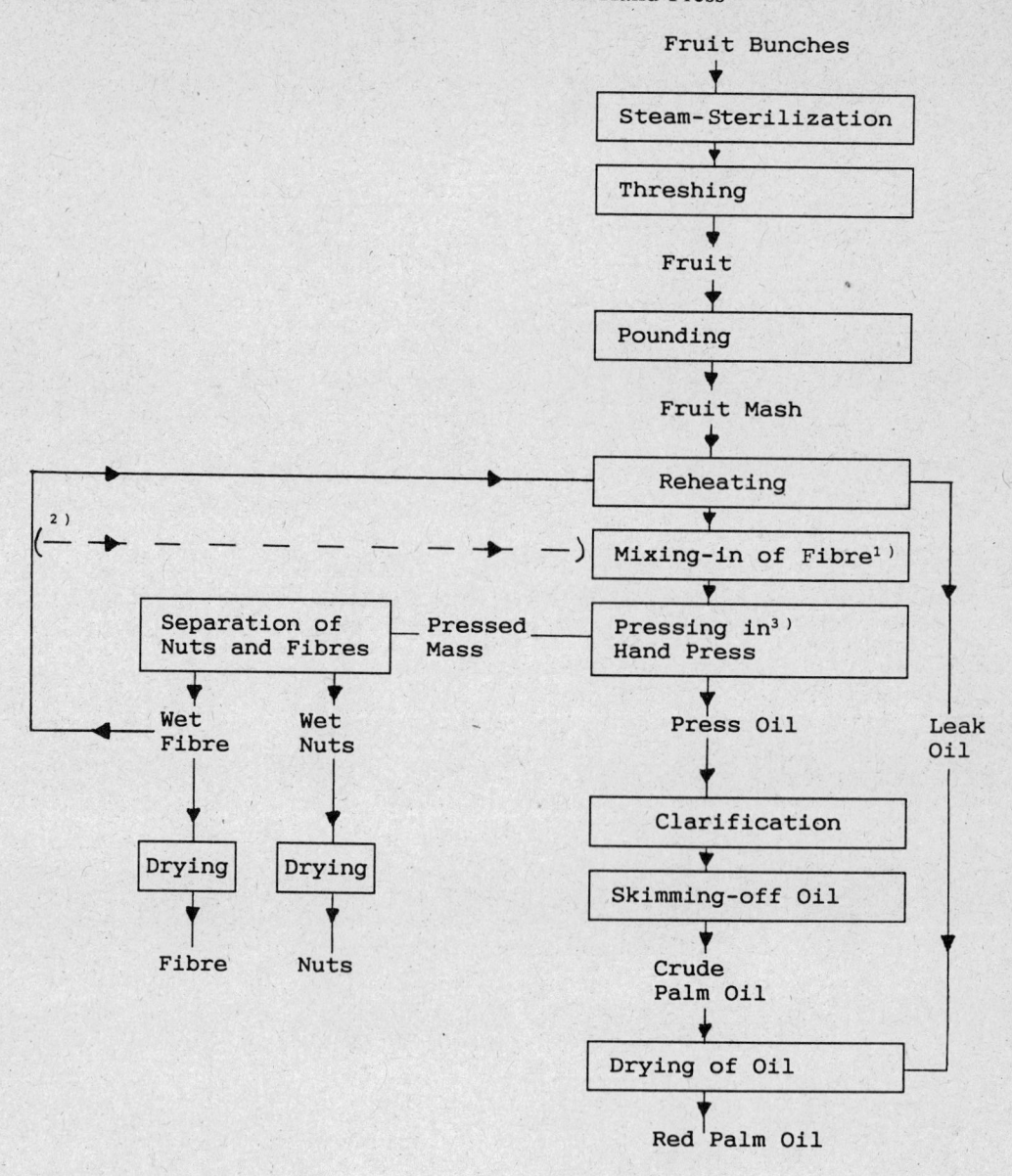

```
                                    Fruit Bunches
                                         │
                                         ▼
                                 ┌─────────────────────┐
                                 │ Steam-Sterilization │
                                 └─────────────────────┘
                                         │
                                         ▼
                                 ┌─────────────────────┐
                                 │      Threshing      │
                                 └─────────────────────┘
                                         │
                                         ▼
                                       Fruit
                                         │
                                         ▼
                                 ┌─────────────────────┐
                                 │      Pounding       │
                                 └─────────────────────┘
                                         │
                                         ▼
                                    Fruit Mash
                                         │
   ┌────────────────────────────────────►▼
   │                             ┌─────────────────────┐
   │                             │     Reheating       │────────────┐
   │  2)                         └─────────────────────┘            │
   │ (─── ─ ─ ─ ─ ─ ─ ─► ─ )            │                           │
   │                                    ▼                           │
   │                             ┌─────────────────────┐            │
   │                             │ Mixing-in of Fibre¹)│            │
   │                             └─────────────────────┘            │
   │  ┌──────────────────┐  Pressed     │                           │
   │  │  Separation of   │───Mass────   │                           │
   │  │ Nuts and Fibres  │   ┌──────────▼─────────┐                 │
   │  └──────────────────┘   │    Pressing in³)   │                 │
   │         ▲     │     │   │     Hand Press     │                 │
   └─────────┘     ▼     ▼   └─────────┬──────────┘                 │
               Wet     Wet             │                       Leak │
              Fibre   Nuts         Press Oil                    Oil │
                │     │                │                            │
                ▼     ▼                ▼                             │
            ┌───────┐┌───────┐  ┌──────────────┐                    │
            │Drying ││Drying │  │ Clarification│                    │
            └───────┘└───────┘  └──────┬───────┘                    │
                │       │              ▼                            │
                ▼       ▼       ┌──────────────────┐                │
              Fibre    Nuts     │ Skimming-off Oil │                │
                                └──────┬───────────┘                │
                                       ▼                            │
                                     Crude                          │
                                   Palm Oil                         │
                                       │                            │
                                       ▼                            │
                                ┌──────────────┐                    │
                                │Drying of Oil │◄───────────────────┘
                                └──────┬───────┘
                                       ▼
                                 Red Palm Oil
```

¹ When processing dura oil palm fruit it is advisable to reheat the recirculated fibre. This is, however, sometimes very complicated. In that case the fibre can be mixed in immediately before pressing.
When processing Tenera oil palm fruit reheating is required in case the fibre (or fibre/nut mixture) is reprocessed.

² When processing dura oil palm fruit, recirculation of fibre is required to prevent the nuts from touching.

³ To be carried out whith any press, sturdy enough to press mixtures of nuts and fibre.

is possible to keep apart small quantities to be processed separately; an advantage, because generally women do not want to have their own fruit mixed with that of others.

Details on an oil palm processing project in Togo using this system are given in Chapter 3.

Although the KIT process can improve much upon the oil recovery, this process is not appreciated everywhere. Some reasons are:
– the large quantity of fibre to be recirculated in the case of fruit with an extremely low pulp content;
– the loss of the possibility to obtain a valuable sludge to be used for food.

To overcome these drawbacks, while still improving workload and oil recovery, a semi-traditional process was introduced for instance in The Gambia and Guiné Bissau. It includes the traditional separation of fibre, oil-containing cream and nuts in water. The fibrous material however is subsequently steam-heated and pressed. The cream is boiled for oil and sludge, as traditionally.

Mechanized systems

Mechanized systems become feasible when really large quantities are involved and regular processing is possible. The first step to be mechanized is the pounding. TCC has developed a horizontal mechanically-driven pounding machine (see Figure 10). This machine is continuously operated by feeding cooked fruit at one side. Digested fruit, ready for pressing, is produced at the other. The capacity of the pounding machine is 100 kg per hour. TCC supplies a steam sterilization kettle and a clarification kettle as well.

The nominal capacity of the TCC system, equipped with a steam sterilization kettle, a pounding machine and two hand presses is 600 kg tenera fruit per day (giving about 144 kg or 24 % of oil). It can be estimated that at least 8 women are required, the equivalent of 8×8 working hours.

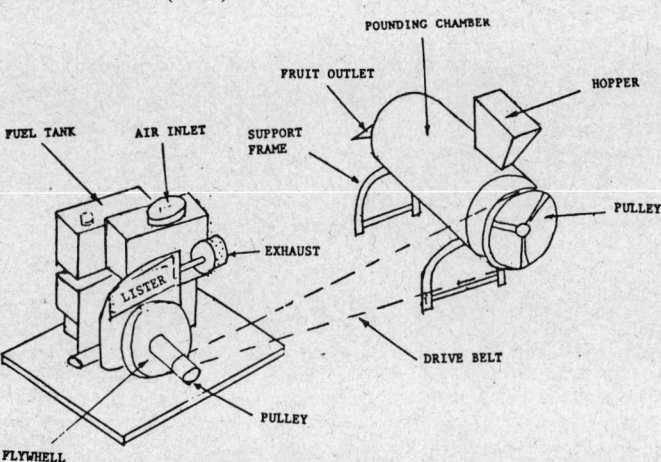

Even more than 40 years ago, the French firm „Pressoirs Colin" developed a continuously working press that carried out the actions of digesting and pressing at the same time. This press was originally meant to be operated by two men. How-

Figure 10: Palm Fruit Pounder (TCC)

ever, it has appeared to be far too heavy for continuous manual operation. When engine driven, this press is an interesting possibility, particularly for the processing of tenera fruit on a relatively large scale. In Cameroon, APICA developed a press, based on the same principles. This press is called the „CALTECH" (see Figure 11).

APICA started with a manual version. This press appeared much too heavy as well. It could not be operated by the same men for more than half an hour. Its nominal capacity is 100 kg per hour. The motorized version (2.3 hp = 1.7 kW) has a capacity of 200 kg per hour. The motorized version of the „COLIN" press, presently on the market as the press SPEICHIM M-10 (4.5 hp = 3.3 kW) has a nominal capacity of 300 kg per hour.

Oil recoveries, reported for the traditional process, the KIT process, the CALTECH and the COLIN expeller, are given in Table 8.

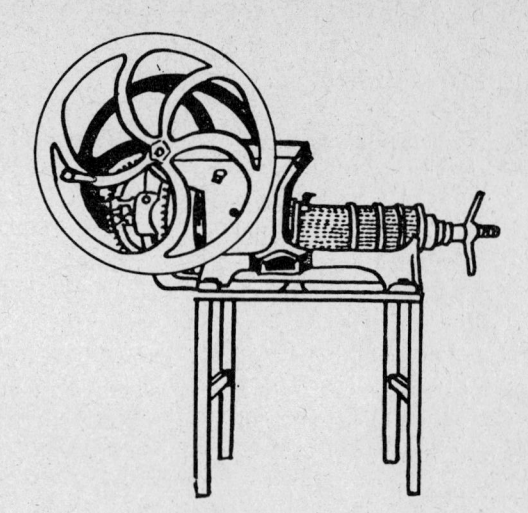

Figure 11: CALTECH Oil Press (APICA) (manual version)

2.2.2 Oil seeds

The service mill

In West Africa, oil seeds are generally processed to sell the product on the week-

Table 8: **Palm Oil: Oil Recoveries Obtained with Different Processes and Equipment**

Variety	Processing system or equipment	Oil recovery	
		% on fruit	% on bunches
Dura	Traditional[1]	1.3	7.2
	CALTECH, hand-operated[2]	(11.3)[4]	7.2
	COLIN, motorized[1]	15.6	9.9
	KIT, Cameroon[2]	(13.8)	8.8
	KIT, Tsévié, Togo: A[3]	13.8	(8.8)
	KIT, Tsévié, Togo: B[3]	15.5	(9.9)
Tenera	Traditional[1]	21.7	13.8
	CALTECH, hand-operated[2]	(25.3)	16.2
	CALTECH, motorized[2]	(26.4)	16.9
	COLIN, motorized[1]	33.0	21.1
	KIT, Agou Yiboe, Togo[3]	(28.0)	17.9

[1] Source: Traut, G. Use of Colin Press for Traditional Palm Oil Processing. In: Report of the First African Small-Scale Palm Oil Processing Workshop, NIFOR, Benin City, Nigeria, 12–16 October 1981, FAO, Rome 1982.
[2] Source: UNATA, Report on comparative tests for COPROCO, 1986.
[3] Source: KIT, Report on feasibility tests for CONGAT, 1986.
[4] ( ) = Calculated on the basis of fruit to bunches ratio = 0.64

ly market. In that case it is a great advantage, if a reasonable quantity can be processed at any one time. This is possible by using the services of the local grain mill, which operates on the principle of a rotating disc. As discussed above, the crushing stage in the traditional process is the most labour-intensive and exhausting one. A women processor can have her oilseed crushed for a fee, if a grain mill is available and if the miller is prepared to crush the oilseed concerned (which is not always the case, because the mill becomes dirty with sticky material). A drawback is of course that a great deal of the added value has to be paid to the miller.

A typical example is the groundnut processing into oil and tunkusa (see 2.1.2). It was observed that a quantity of 18.1 kg shelled groundnuts was processed into 3.7 kg oil and 15.7 kg tunkusa (giving 14.6 kg kuli-kuli) by 3 women in only 8 working hours, using the service mill for grinding the groundnuts. Crushing by hand would have required about 20 hours alone. Service mills of the type suitable for grinding oilseed can be found over the whole of West Africa. Oil seeds which can be crushed in such mills include: groundnuts, palm kernels, and shea nuts. Coconut can be grated by service mills equipped with a grater.

The use of mechanized size reduction replaces not only heavy and exhausting work, but also improves oil recovery, as a much larger number of oil-containing cells are opened to produce oil.

The ghani

On the Indian subcontinent, one will come across the ghani oil mill as the village level processing system.

The ghani consists of a mortar and pestle in which the seed is crushed. Pulverization and oil expression are carried out at the same time by rubbing the seed between the pestle and the wall of the mortar. Water has to be added to realize a preparation process, called „cooking", by which the oil emulsion in the cells is broken, and the micro oil droplets are combined into larger ones. The ghani process requires much mechanical energy. A ghani operated by one bullock (the equivalent of 0.35 kW) can process 5 kg oilseed in about one hour. Hence, 0.35/5 or 0.07 kWh are required to process 1 kg of an oilseed into oil. This energy consumption is about equal to the maximum amount of energy required by small oil expellers.

(Oil expellers require between 1.5 and 2.5 kW to process 35 kg of an oilseed per hour, or between 0.04 and 0.07 kWh per kg oilseed.)

Figure 12: Power Ghani

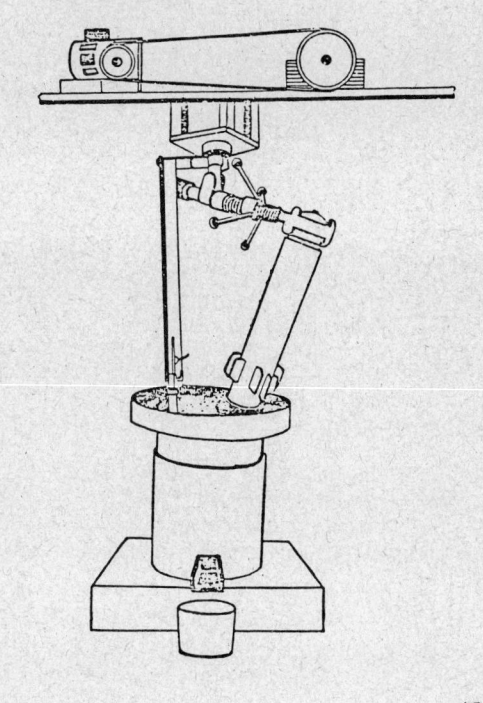

Besides, the ghani requires a lot of maintenance and repair, as the surfaces of the mortar and the pestle suffer much abrasion. The mortar can be recut 2-3 times but has then to be replaced. The pestle needs to be replaced at regular intervals as well. Every 48 hours these surfaces should be checked on wear. Also the bearing at the top has to be checked regularly.

The introduction of a mechanized version into Tanzania, the so-called „power ghani" has had disappointing results, since special skills seem to be required for their operation and maintenance. In India, these mechanized versions are quickly replacing the original animal-driven ones. Also modernized versions are available, equipped with a turning mortar instead of a turning pestle as is the case with the traditional ghani. Ghani oil is highly appreciated because of its special flavour, particularly in the case of mustard seed oil.

The hand press

Although the wet process can be improved by mechanized size reduction using a motorized disc-mill or a motorized grater, it can be of advantage to introduce the dry process, using a hand press.

As no motorized equipment is required, it is particularly suitable as a women's activity, enabling the maximum of the value added to be retained by the processors themselves. The dry process consists of the five process-steps shown in flowsheet 5.

KIT has developed unsophisticated equipment to carry out the dry process, completely by hand. It can be used to process oil seeds at the village level as: groundnuts, sunflowerseed, palm kernels, coconuts, sesame seed, rape- seed, castor seed and shea nuts.

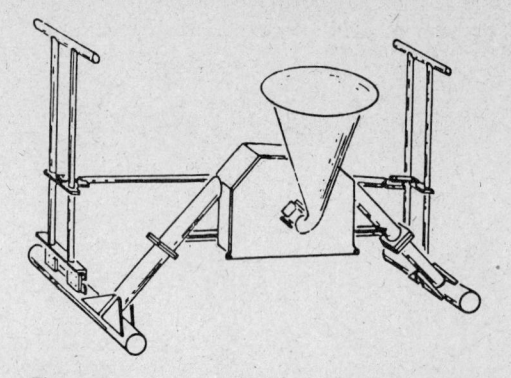

Figure 13: Palm Nut Cracker (KIT/UNATA)

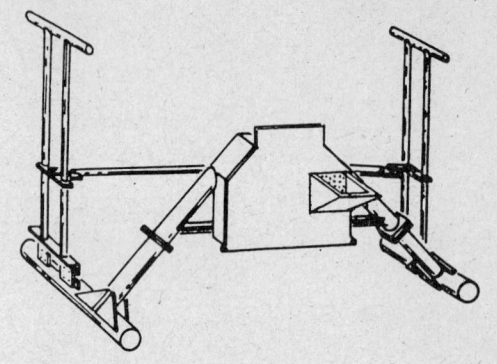

Figure 14: Cocos Grater (KIT)

Figure 15: Roller Mill (KIT/UNATA)

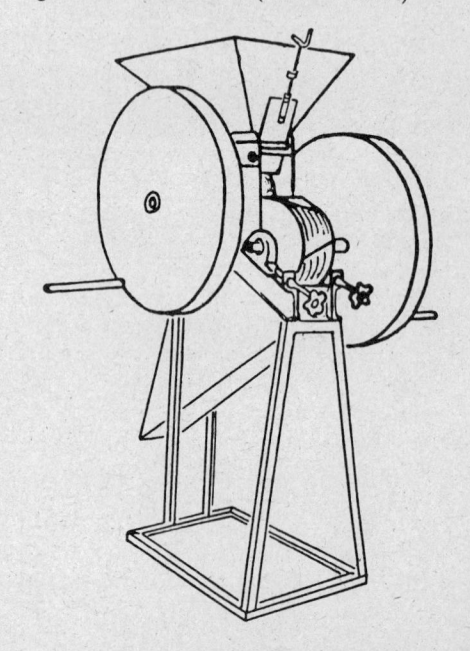

Flowsheet 5 Dry Process for Processing Oil seeds (General Flowsheet)

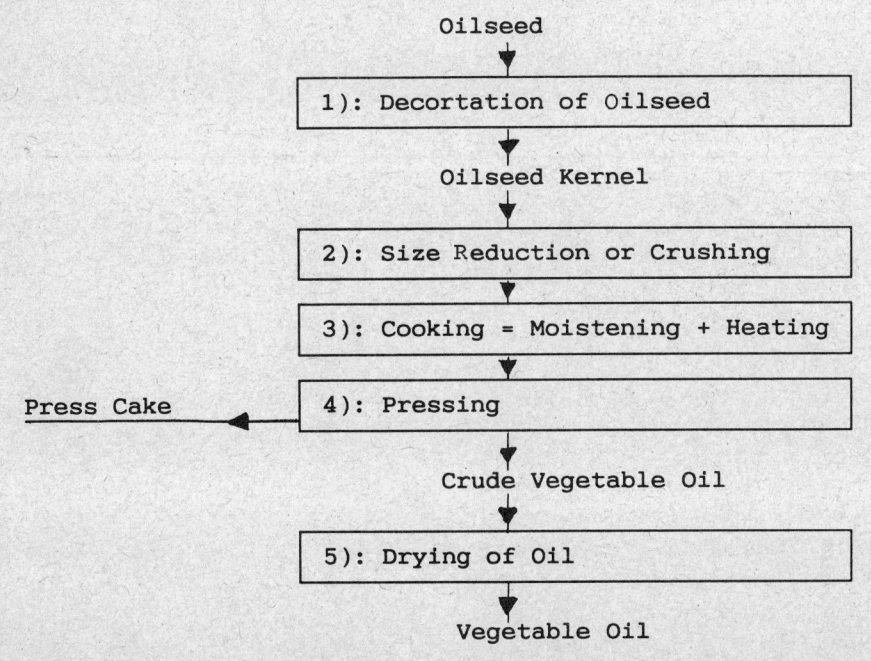

```
                        Oilseed
                           ▼
            ┌──────────────────────────────┐
            │ 1): Decortation of Oilseed   │
            └──────────────────────────────┘
                           ▼
                     Oilseed Kernel
                           ▼
            ┌──────────────────────────────┐
            │ 2): Size Reduction or Crushing│
            └──────────────────────────────┘
                           ▼
            ┌──────────────────────────────┐
            │ 3): Cooking = Moistening + Heating │
            └──────────────────────────────┘
                           ▼
Press Cake ◄────────┤ 4): Pressing │
            └──────────────────────────────┘
                           ▼
                   Crude Vegetable Oil
                           ▼
            ┌──────────────────────────────┐
            │ 5): Drying of Oil            │
            └──────────────────────────────┘
                           ▼
                     Vegetable Oil
```

1) **decortication:** to prepare a rawmaterial for further processing with the highest possible oil content,
2) **size reduction:** to obtain a well crushed material, with still some coherence, to facilitate filtration during pressing but to avoid the pressing-out of fine material with the oil. The best way to crush is between rollers into very fine flakes, preferably thinner than 0.1 mm. Particles to start with should be smaller than 5 mm diameter.
3) **cooking:** to prepare a mass in which the oil is present in a form that it can relatively easily be pressed out. During cooking the following processes take place:
   - weakening of the cell walls,
   - enaturing of proteins, destabilizing the oil emulsion, the original form in which the oildroplets are present in the cells,
   - coalescence (flowing together) of micro oil droplets into larger ones,
   - diminishing of viscosity because of higher temperature.
   Cooking is a process in which temperature, moisture and time play an important role.
4) **pressing:** to separate the oil from the rest of the seed. To be able to use a press with the lowest possible maximum pressure, size reduction and cooking should be carried out with great care. For most oil seeds optimal conditions are achieved by pressing at about the same moisture content as that of the original rawmaterial. In that case, the maximum pressure should be 60 kg/cm². Higher pressures have, in that case, little effect, as the mass will be extruded through the holes in the press cage. Oil recoveries are good, while equipment remains simple. At higher maximum pressures either cage contents become inefficiently small or the equipment becomes much too complicated and too expensive.
5) **drying** of the oil: to prepare a dry and pure oil, fit for long-term storage. Oil should be dry and free of impurities. The oil has therefore to be heated to 130°C to remove all traces of moisture. After leaving the oil to stand for a few days, the impurities will have settled and pure oil can be decanted. The oil should be filled to the top in clean and dry bottles or tins, and stored in a dark and cool place. Oil treated in this way has a shelf-life of at least 6 months.

Figure 16: Heating Oven (KIT)

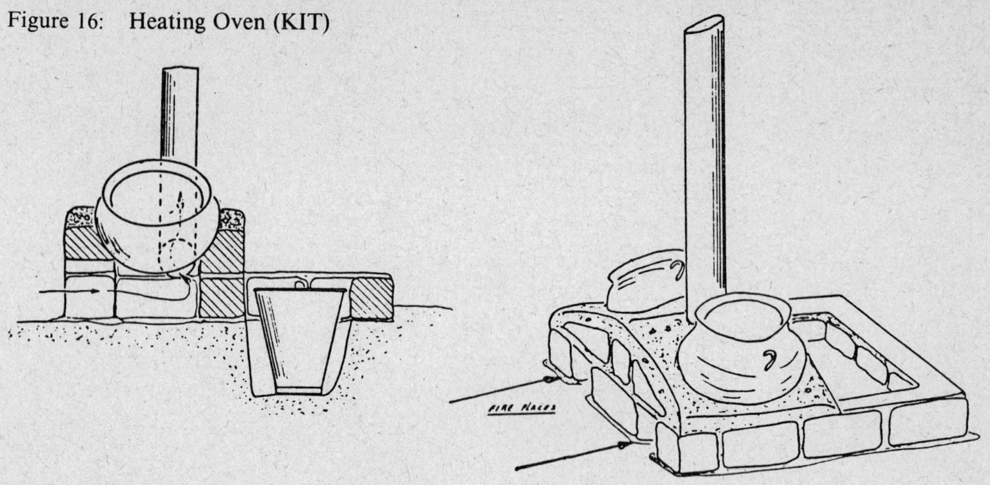

The equipment consists of the following (hand-operated) machines:
- sunflowerseed decorticator or palm nut cracker (see Figure 13), to be rebuilt into a hammermill
- winnower
- roller mill (see Figure 15)
- „cooking" furnace (see Figure 16)
- hand press (see Figure 17)

Table 9: **Typical Performance of KIT/UNATA Hand-operated Equipment**[1, 2]

Raw Material (R.M.)[3]	Preparation	Capacity kg/day	Oil-content of R.M. %	Oil recovery on R.M. %
Groundnuts	decorticated crushed cooked	80	45	33–35
Sunflowerseed	partly decorticated crushed cooked	150	32–35	19–21
Palm kernels	broken crushed cooked	depending on milling capacity	47.5	33–35
Coconuts	meat partly dried[4] grated cooked	350 nuts	10.2 kg	7.5 kg per 100 nuts[5]

[1] Other seeds that can be processed following the KIT/UNATA hand-operated system include sesame seed, rape and castor seed.
[2] Performance according to field experience.
[3] Seed as described: groundnuts (decorticated), sunflowerseed (not decorticated), palm kernels (after cracking of palm nuts), coconuts (as such).
[4] Dried for two days (=30% moisture).
[5] Oil content and recovery are not given as a percentage, but as kg per 100 nuts!

Figure 17: Spindle Press (UNATA)

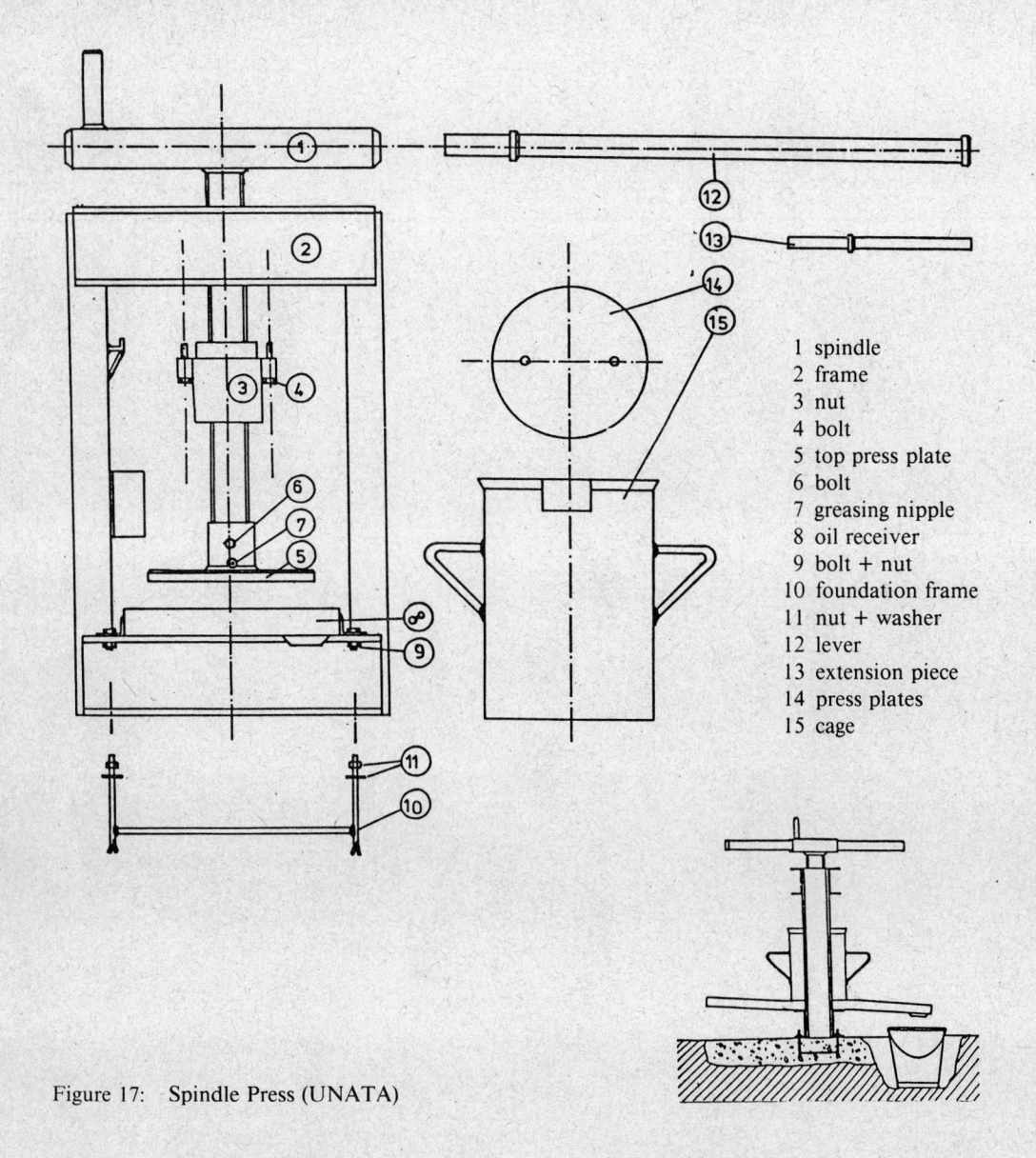

1 spindle
2 frame
3 nut
4 bolt
5 top press plate
6 bolt
7 greasing nipple
8 oil receiver
9 bolt + nut
10 foundation frame
11 nut + washer
12 lever
13 extension piece
14 press plates
15 cage

A grater for grating fresh coconut (a drill type) and a different type for half dried copra (disc type, see Figure 14) are under development. Originally, hydraulic hand presses were used, based on hydraulic lorry jacks.

It appeared, however, that already at a maximum pressure of 60 kg/cm² good oil recoveries could be obtained and therefore the UNATA spindle press was adopted. The press was redesigned to meet the requirements and has been put into production (see Figure 17). Only for shea nut processing does this press not give a good oil recovery, as a maximum pressure of 120 kg/cm² is required. Up to now, only the hydraulic press, designed

49

for the GTZ/GATE project in Mali, can be efficiently used to process shea nuts.

However, a spindle press for shea nut processing is under development. The performance of the KIT process, using the UNATA spindle press, is given in Table 9.

In Tanzania, IPI has developed equipment for processing sunflowerseed, following the same process. The equipment is described in Chapter 5.1.2.

IPI has carried out detailed research on the effect of the different process steps described above on the overall press efficiency; i.e. the oil recovered as a percentage of the oil present in the rawmaterial. The results are illustrated in Figure 18.

For its hand-operated system, IPI advises boiling the crude oil with salt (2 %) and water (10 %), decanting and filtering through a cloth. The performance of the IPI system is summarized in Table 10.
In Chapter 3, details are given on the sunflowerseed processing project in Zambia and on the GTZ/GATE project in Mali. The economic aspects of these activities are worked out in Chapter 4.

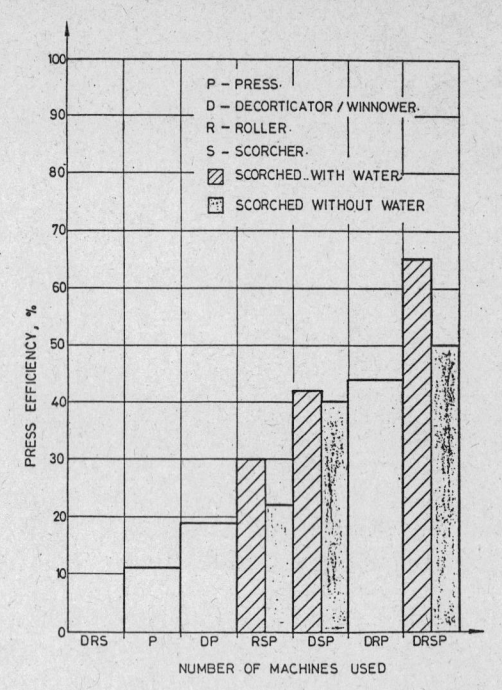

Figure 18: Oil Yield as Determined by the Number of Machines Used in the Process for Sunflower Seeds (IPI)

Legend for combinations tested:
- P: Press (alone)
- DP: Decorticator + Press
- RSP: Roller mill + Scorcher (cooking furnace) + Press
- DSP: Decorticator + Scorcher + Press
- DRSP: Decorticator + Roller mill + Scorcher + Press

Oil-Expellers

The dry process cannot only be executed by mechanical or hydraulic presses, but also by continuously operating screw presses, generally named oil expellers. An oil expeller consists of a perforated cage in which a tapered screw turns. The screw is

Table 10: **Typical Performance of IPI Hand-operated System**

Raw Material (R.M.)[3]	Preparation	Capacity kg/day	Oil-content of R.M. %	Oil recovery on R.M. %
Sunflowerseed	partly decorticated crushed cooked	210		19

Table 11: **Typical Performance of Oil Expeller MINI 40**[1,2]

Seed	Preparation	Capacity kg/hour	Oil-content of		Oil recovery from seed %
			seed %	cake %	
Sunflower	not decorticated; cold	35/40	35.0	15/18	18–22
Palmkernels	ground 2.4 mm cold	35/40	47.5	20	34
Copra	broken 10 mm	40	65.0	30	50
Rape	cold	35/40	42.0	17/20	28

[1] Manufactured by Simon-Rosedowns, Hull, England.
[2] Figures from the manufacturer for one pressing only. More oil can be produced by pressing a second time.

tapered in a way that the free space between the centre of the screw and the cage gradually becomes smaller to the end of the cage. With this system, very high pressures can be exerted on the material to be pressed.

The rawmaterial has to be prepared in principle by decorticating, crushing between rollers and cooking. Depending on the type of machine, however, some oil seeds can be processed even when not decorticated. Expellers are usually driven by petrol or diesel engines or electric motors but can also be run on animal or water power.

The performance of an expeller developed especially for processing on a small scale is presented in Table 11. Other small expellers give comparable results.

The performance of an expeller developed for processing in one step (so-called „deep" pressing) is given in Table 12.

In the above tables, all data on performances are those given by the manufac-

Table 12: **Typical Performance of Oil Expeller MRN (AP VII)**[1,2]

Seed	Preparation	Capacity kg/hour	Oil-content of		Oilrecovery from seed %
			seed %	cake %	
Groundnuts	10% husks added; cold	70	45.0	8–10	39
Sunflower	not decorticated; cold	70	35.0	8	29
Palmkernels	broken; cold	60	47.5	10–12	41
Copra	broken;	85	65.0	8–10	62
Rape	cold	90	42.0	10	36
Soyabeans	cold	70	20.0	8	13

[1] Manufactured by Maschinenfabrik Reinartz, Neuss, F.R.G.
[2] Figures from the manufacturer. Even higher recoveries can be obtained with proper preparation including heating.

turers and may have been determined under favourable conditions. After some time of intensive use, the oil recovery of expellers usually declines until an overhaul becomes necessary. Furthermore, it should be kept in mind, that oil expellers require considerable maintenance and repair, for which expensive spare parts are required. Generally, conditions in villages are such that it is very difficult to run expellers economically, if they can be kept running at all! Expellers are generally found in towns, e.g. in East Africa, where they are sometimes made available as service mills.

More details on expellers as alternative oilseed processing equipment are given in Chapters 3, 4 and 5.

2.3 District level

Oil processing at the district level (in the sense of a group of a few villages), offers interesting possibilities.
At this level, however, the technical performance of the equipment is only one side of the picture and, in fact, less problematic than the management of such a project. Important aspects include:

– the ability of the people concerned to organize themselves (in a cooperative or in a private business),
– the ability to handle funds,
– the ability to take care of the rawmaterial supply and
– the marketing of the products.

Nevertheless, centralization can contribute considerably to the feasibility of the more sophisticated technology as already described for the village level. For instance in the case of *oil palm fruit processing*, mechanized equipment, such as the TCC pounding machine and certainly the CALTECH and COLIN expellers, need to be well utilized owing to the high investments involved. Where the infrastructure is well developed and the distances not too far for economic transportation, a combination of the raw material resources of several villages and a centralized processing facility could be a realistic alternative to processing at the village level.

In that case, one could think of a well engineered unit, equipped with:

– steaming facilities for bunches and loose fruit
– threshing facilities
– a good quality expeller type press (such as the CALTECH or COLIN)
– clarification tank
– oil dryer.

In the case of *oilseed processing*, an expeller — often to be imported — could become a possibility, provided that technical prerequisites are fulfilled, such as the availability of spare parts and the necessary skills for maintenance and repair.

Apart from the technical aspects, it should be kept in mind that such highly mechanized technologies are in principle:

– capital intensive,
– labour-saving,
– economically sensitive to bad harvests and falling oil prices, and
– socially geared to the use by men instead of women.

A considerable decline in employment opportunities at the village level (particularly for the women) might therefore be the effect of a larger scale oil processing operation at the district level.

Although such units might look attractive from the technical point of view (see details in Chapter 5), the setting up of centralized units cannot be recommended as long as there are still doubts as to the possibility of finding appropriate solutions for the technical as well as the management and more human-oriented problems.

3. Case Studies

3.1 Shea nut processing by women in Mali

Background

Since 1982, a GTZ/GATE/DMA project has been engaged in Mali to develop and disseminate a system for processing shea nuts that improves on the traditional system. For this project, an oven, a hydraulic hand press and a cake-expel stand have been developed by KIT. After introductory tests, the local production of the equipment has been set up and the equipment is being disseminated.

Organization and management at the village level

In villages, where shea nut trees grow, practically all the women collect and process shea nuts. The collected nuts become the personal property of the collector and are stored in her own pits. The processing is traditionally carried out at the level of the extended family by a group of women, who assist the collector. Occasionally small quantities are processed by individuals.

Adaptation of the shea nut press means that the processing has to be carried out at the village level. As the village women are generally organized and headed by a president, who is assisted by a group of elder women, this is quite feasible. However, first of all, the men, headed by the village chief, have to advise. The installation of a press is a village affair and has to be approved and backed by the men. The contribution by the village (50 % of the total investment costs for equipment and building) is raised by the men and the women.

The president and some representatives of the women supervise and organize the utilization of the press. They are also the owners of the press. A treasurer has to collect a small fee from the owner of the nuts in proportion to the quantity processed; money which is to be kept separate. It serves as a reserve fund to cover expenses for maintenance, repair and amortization. To have the fund administered in a formal way, a literate person should be available. Otherwise, a more informal administrative system should be implemented.

The press is installed under supervision of the GTZ/GATE project. The group appoints a few women (some older, some younger) to be trained in the operation and maintenance of the press. These women assist the processors and will train them in turn. The way in which the press is utilized is essentially a service system, by which the press is made available to the women of the village for a fee. The owner of the nuts is responsible for the processing and must provide the operating crew.

Process and equipment

The nuts to be processed are removed from the pit, pre-dried in the sun and decorticated. The kernels are subsequently pounded into a powder (with particles smaller than 5 mm). If the powder is dry, some water (about 10 %) should be added and left to be absorbed for an hour at least. The mass is divided into 5 kg batches and heated in a pot over a fire (to about 120°C) and kept hot for about 1 hour in the oven. The hot mass is poured into a preheated cage and pressed. When pouring, the mass should be divided in small portions of about 1 kg by means of a pressplate. After pressing, the cage is removed from the press and put on the expel stand to expel the cake, after which the oil extraction process is repeated, using the oil cake as raw- material. For this second pressing, the cake should be pounded and sieved (to be sure that all particles are smaller than 2 mm) and water (about 10 %) should be added. Finally, the extracted oil, containing some brown particles, is boiled with a little water and some juice of okra and lemon, to obtain a clear, white oil. The oil is left to cool down to form a solid fat, called shea butter. The cake is used as fuel.

The equipment required consists of:
– an oven, made up of a fire-place with a pot (no. 15, contents 40 l) and a heated box to hold a few buckets, covered with lids,
– a hydraulic hand press (based on a 30-ton lorry jack and equipped with two press cages with pressplates),
– a cake expel stand (to remove the oil cake from the cages) and
– some minor equipment like mortar and pestles.

The cost of the press with expel stand is F CFA 400 000, for the oven F CFA 40 000 and for the minor equipment around F CFA 10 000 (1987 prices). F CFA 50 are equivalent to French F 1.0, which is approximately US $ 0.15.

Figure 19: Shea Nut Processing in GTZ/GATE Project

Results and experience

Total working time required for processing 10 kg shea kernels is about 10 man hours, of which about 3 hours are required for the pounding. A group of five women, which is a common case, would process some 50 kg of kernels per working day on average. Experienced groups can process up to 80 kg per day. Oil recovery is, depending on the quality of the nuts, between 35 % and 42 % on dry kernels, sometimes even higher.

Women are pleased to use the press, because:
1. heating of the nuts in a traditional oven for about three days at least can be omitted, which saves a lot of firewood and time;
2. the mass is easier to pound, while the crushing between stones can be omitted, which makes the work much faster and lighter and saves therefore time and energy;
3. very little water is required, which saves a lot of work;
4. the oil recovery, which of course depends on the quality of the nuts, is higher than traditionally (an estimate is about 13.5 % on dry kernels; i.e. 38.5 % average yield against 25 % in the traditional process);
5. the cake, which traditionally is discarded, can be used as fuel.

Another positive aspect is that this technology can be mastered by women and that the men are not really interested to take it over. The most probable reason for this is that the crucial steps, such as „size reduction" and „cooking", are still carried out in the traditional way, by pounding and a pot-over-a-fire respectively.

Shea nut processing is mainly carried out in the dry season, when the harvest is over. However, if money is needed and the rawmaterial is available, the women process oil in August as well. In a village in South Mali, 27 women are recorded as using the press and 1 350 kg nuts were pressed in one year (an average of 50 kg per woman per year). In the same village, women collected other oil seeds as well as Niam Seeds *(Lophira alata)* for food-oil, Physic Nut *(Jatropha Curcas)* and seeds of the wild olive *(Ximenia americana)* for soap-oil. This can be seen as a first attempt towards a diversified production and will have an additional positive effect on the amortization of the equipment.

Although the women were trained with care, many women had difficulties with technical aspects; particularly with the hydraulic jack equipment. In some cases, the release valve was completely unscrewed and the hydraulic oil was lost. Besides, the women had to develop a sympathetic understanding for the equipment in order to prevent damage, for instance by pumping long strokes at a steady rate to conserve the pump and to stop at a certain maximum pressure to protect the press and save energy.

Also, an understanding of the process required time. Refresher courses are needed in order to keep the operation up to the required standard. Regular maintenance appeared to be required as well. Young villagers and some women were trained to carry out small maintenance and repair jobs such as topping-up with hydraulic oil or renewing a pump seal.

Technically, the press appeared to be not without its problems. The jack pump showed not only rapid wear at its seal, but also of its mechanical parts. On the other hand it appeared that the jack was not sufficiently well supported by the frame, thus causing the base of the jack to break. An improved pump for the jack has recently been designed. Also the frame will be reinforced. It is expected that these improvements will have positive effects. Nevertheless, due to the advantages of the new processing system and the start-up of a service and maintenance programme, the demand for the shea nut presses is steadily increasing. By early 1987, the GTZ/GATE/KIT shea nut processing equipment had been installed in 35 villages in Mali. In addition, enquiries from neighbouring countries are increasing in number. Recently, the new technology has been introduced to Burkina Faso after technicians and extension workers had been trained in Mali.

Finance

It cannot be expected that the cost of investment of F CFA 440 000 for the technical items (oven parts, hydraulic press and expel stand) can be paid in full by any village. Therefore, the equipment is normally partly subsidized by the project or other international organizations. The village usually is expected to build the accommodation (for approximately F CFA 100 000) for the press and to raise a certain contribution for the equipment. A small fee is asked from the women (e.g. F CFA 25 per 5 kg bucket) for a reserve fund to finance maintenance, repair and amortization.

Alternative possibilities

For improving and facilitating the processing of shea nuts, an alternative solution is the use of a motorized grain mill. Because the processing of shea nuts makes the discs of the mill very sticky (requiring an extra cleaning procedure), many village millers are reluctant to make their grain mill available for the custom milling of shea nuts or alternatively demand much higher fees. If a motorized grain mill is available to the village, it is recommended to install a separate mill that can be driven by the same engine.

In Chapter 4, a financial appraisal is given based on the average price for nuts of F CFA 30 and for shea nut butter of F CFA 300 per kg, as prevailing in early 1987 in the rural areas of Mali. At that time, the local cost for milling of grain by a service mill was about F CFA 25 per kg.

Local production of equipment

To be able to start the distribution of the presses, local production had to be set up.

At first an existing construction company was given the opportunity to take up the production of the presses, guided by an expert from KIT. As it appeared that this company only wanted to continue if they could charge a 100 % profit, it was decided to establish a special production unit. A small partnership of craftsmen was advised by the GTZ/GATE project in establishing its own workshop with the press as the most important product. At present, this workshop manufactures the presses, supplies the spare parts and repairs the equipment, including the hydraulic jack. However, the supply of imported items, such as jacks and their spare parts, still gives problems. A reliable import line has yet to be organized.

3.2 Hand-operated sunflowerseed processing in Zambia

Background

Since 1985, a project has been carried out in Zambia to establish a few fully hand-operated sunflowerseed processing units, mainly financed by the government of the Netherlands. The objective of the project is to test the feasibility of hand-operated sunflowerseed processing in Zambia. As quite a few projects are interested in assisting the establishment of such units, a preparatory familiarization programme has already been started by the Technology Development and Advisory Unit (TDAU) of the University of Zambia (UNZA), Lusaka Campus. The equipment will be manufactured locally.

The test-units were established at the Kasisi Mission near Lusaka, at the Kaoma-TBZ Scheme and in the Gweembe Valley.

Organization and management at the unit level

The unit established at the Kasisi Mission is an example of a quite formal enterprise. The unit is an external activity of the Training Centre to whom it is responsible. The Training Centre as a whole is the responsibility of the fathers of the Kasisi Mission. The unit is operated by a group of six school-leavers, headed by a group leader. They work the normal working hours under the supervision of the Centre and are paid as agricultural labourers. The Centre constructed the building and provided the working capital. The equipment has been imported and was financed by outside assistance. Financial management is taken care of by the Mission and the seed is bought in the period just after harvest. All the seed required until the new season is bought before the prices for the new season are announced. The seed is stored at the Centre.

At the Kaoma-TBZ Scheme and in the Gweembe Valley, the units are operated by women's groups. At Kaoma-TBZ, the Kweseka women's group is responsible for the operation of the unit. Guidance is provided by the extension staff of the Ministry of Agriculture who made some working capital available for the procurement of seed.

In the Gweembe Valley, women's groups are responsible as well and are guided by staff of a project by the Gossner Mission, who provides the seed.

The Kasisi unit is well established. Its operation is described below as a typical example of how a well managed unit can operate.

Initially, a processing capacity of 3 bags or 150 kg per day was reached, but oil recovery was very low. This low oil recovery was caused, for example, by:

– improper dehulling and large losses of kernel material during this step,
– improper moistening, by not letting the water be absorbed for one hour,
– improper pressing, by not waiting long enough for all the oil to become released.

In May 1986, refresher training sessions given by the Kasisi mission father responsible for the centre, revealed that proper processing, carried out with discipline and using good seed, gave much better results as follows:
– early planted local seed (composite 75), with an oil content of 35.5 % gave 9.8 kg (or 10.3 l) oil per bag (19.6 % recovery on seed),
– late planted hybrid seed, with 37.0 % oil gave 11.4 kg (or 12.0 l) oil per bag (24 % recovery on seed).

In September 1986, an evaluation showed that these figures were being realized on a continuous basis. A point that still needed special attention, however, was the adjustment of the clearance between the rollers of the roller mill.

The set-up at Kasisi works very well, mainly because of good leadership and management. By refresher sessions, technical problems with the process could be overcome, while maintenance and repair was easily cared for by the mission workshop.

The women's groups at Kaoma TBZ and in the Gweembe valley are lacking leadership and management. Also, refresher sessions are necessary to maintain standards of processing. The organizational system, by which the problems of these groups could be overcome, has still to be worked out.

Figure 20: Sunflower Seed Processing in Zambia

Process and equipment

The process applied is the „dry process", including:

– dehulling of the seed (decorticating and winnowing)
– size reduction (crushing in a roller mill)
– cooking (moistening and heating)
– pressing (in a hydraulic hand press or in a spindle press).

At Kasisi and Kaoma-TBZ, the units are equipped with a CECOCO decorticator and winnower, and a KIT designed roller mill, heating unit and hydraulic hand press. In the Gweembe Valley, KIT designed dehulling equipment replaces the CECOCO equipment, and the UNATA spindle press replaces the hydraulic press.

To prepare for local production, these types of decorticator and winnower are currently being tested at Kasisi. Their performance and durability appear satisfactory. Unfortunately, the hydraulic press gives problems. The hydraulic system seems to be quite vulnerable and difficult to repair. Since the same pressing results can be obtained with a simple spindle press, which is also easier to manufacture and repair locally, it has been proposed to change to the UNATA spindle press.

At Kasisi, the operators start on a normal working day at 7 a.m., divided into three groups of two. Each group is charged with the processing of one bag (or 50 kg) sunflowerseed. The group that has finished the day before with crushing starts with mixing-in the water. After one hour they start heating, followed by pressing. (Between heating and pressing, the mass is kept hot in closed buckets in a kind of a hay-box.) After pressing, this group takes up decorticating followed by crushing to prepare the material for the next day. Another group starts with crushing followed by mixing in water, then heating and pressing. After pressing, they start decorticating to prepare for the next day. The third group starts with decorticating, followed by crushing, mixing in the water, heating and pressing. A complete cycle for 50 kg requires about 6 effective working hours. Some more time is required for preparatory work, rest to recover from heavy manual work and maintenance. The group takes a break at noon. At 2 p.m. the work is resumed until 5 p.m.

At Kaoma-TBZ, the women, when they were still motivated, worked four days a week with ten persons, processing in three groups about 25 kg seed per group per day. Later on, the women's groups became smaller, and late 1986 they processed only 50 kg seed per day, two days a week.

Results and experience

The process was introduced at Kasisi in November 1985. The group was trained for one week. Due to problems of getting a license the production could not start before January 1986. Between January and mid-April the group worked quite independently and with little supervision. In total 11.4 tons of seed were processed and 1665 l sunfloweroil sold. That equals a net production of 7.3 l of oil per 50 kg bag of seed.

Finance

From January to April 1986, the financial results of the Kasisi unit were disappointing, as the income from oil could only cover rawmaterials and wages and no other costs. This was due to:
– low oil recovery (only 7.3 l per bag of 50 kg)
– low production capacity at the beginning (below 150 kg per day)
– increased wages (an increase of about 50 %)
– low oil price (K 5.00 per l, because subsidized oil flooded the market).

For the season 1986/87, the Mission had purchased the required 30 tons of sunflowerseed of good quality at the official price (K 42 per 50 kg bag) just in time, before the new producer prices (of 70 K per 50 kg bag) had been announced. A reasonable daily budget could therefore be made (see below).

The main problem for Kasisi will remain the difficult economic situation in the country, with steeply rising prices for sunflower seed to compensate for rising production costs, and heavily subsidized relatively constant prices of vegetable oil.

Because of the high sunflowerseed prices and the low oil prices, together with their low productivity, oil production in the 1986/87 season seems not to be remunerative enough for the women of the Kweseka group at Kaoma-TBZ. As long as this situation continues, not much production can be expected at Kaoma-TBZ.

In the Gweembe Valley, sunflower seed is a little cheaper and oil a little more expensive, which makes oil production more remunerative. It is not yet known if this margin is interesting enough for the women concerned to take up oilseed processing on a continuous basis.

Expenses per day:	
3 bags sunflowerseed @ K 42/bag :	126.00
wages for 6 agricultural labourers :	32.40
other costs :	10.00
amortization equipment @ K 10/bag :	30.00
amortization building @ K 10/day :	10.00
Total	208.40
Income per day:	
30 l oil at K 7 per l (just below price of subsidized oil, September 1986) :	210.00
cake (used as animal feed by the mission) :	
Net Cash Income per day: :	1.60

Alternative possibilities

Alternatives for fully hand-operated sunflowerseed processing in Zambia are motorized expellers, as presented in Chapter 2. In Chapter 4, the feasibility of a locally manufactured hand-operated system is compared with the feasibility of alternative motorized expeller systems.

The price of a locally manufactured hand-operated system (basis September 1986) is estimated at K 20 000. At that time, 1 K was equivalent to 0.17 US $.

3.3 Oil palm fruit processing as a women's activity in Togo

Background

Early in 1986, the Togolese organization CONGAT requested KIT to introduce its improved system for oil palm fruit processing in Togo. The project was financed by the Dutch NGO: ICCO. In the framework of this programme, a unit for processing Tenera oil palm fruit was established at Agou Yiboe. The objective of the programme was to demonstrate the feasibility of this fully hand-operated system to improve the living conditions of women in the south of Togo.

Organization and management

The CFAE (Centre de Formation Agricole et Economique) at Agou-Yiboe, Togo, guides a group consisting of about 18 young women. In addition to the staff of the CFAE, the group is guided by a Canadian lady volunteer. A unit for processing tenera oil palm fruit has been established at the CFAE to provide the women with productive employment. Bunches of tenera oil palm fruit are made available as the rawmaterial to be processed by the CFAE from their own plantation. The price quoted to the group is the same as the CFAE gets from a nearby palm oil mill.

The group is headed by a chairwoman, who is assisted by a core group of women. Originally, the group met for training sessions on the subject of hygiene, child care and needlework. Later on they started with agricultural work on their own account, raising crops such as maize. As the income earned by the women from agriculture was disappointing, the processing unit has been set up.

CONGAT has made the equipment available and the CFAE the working capital; equipment and working capital have to be reimbursed by the group. The staff of the CFAE assists in the management of the funds.

Process and equipment

The unit has been set up to process 500 kg of tenera oil palm fruit bunches per day, twice a week during 6 months of the year. The unit consists of an open shed and is equipped with the following equipment:

– a large cooking kettle for the steaming of bunches
– a threshing grid
– a concrete pounding mortar
– 4 reheating kettles
– a hand-operated spindle press UNATA 4201
– a clarification kettle
– cooking pots for oil drying.

The processing of oil palm fruit is carried out in the following way: about 500 kg of

61

Figure 21: Oil Palm Fruit Processing in Togo

bunches are bought from the CFAE on the day before the actual processing is to be carried out. They are loaded into the steaming kettle and steamed for about four hours in the afternoon, under the supervision of two women.

The next day starts at 7 a.m., with threshing and pounding. The fruit is still warm, but not too hot to touch. As soon as a reheating drum has been filled, the fruit is reheated. This step can be the most crucial, since, if not carried out correctly, the oil recovery will be disappointingly low. Since the mass of pounded tenera fruit is quite solid and difficult to penetrate by steam, at least three hours are required for proper reheating. After finishing threshing and pounding, carried out by 8 women in about three hours, there is a pause for about one and a half to two hours as one has to wait for the reheating to be completed. Around noon, or a little later, pressing starts after which fibres and nuts are separated and the fibre is collected in an empty reheating drum and reheated again. The press fluid is collected in the clarification drum. The leak-oil, obtained during reheating, is put in an oil drying pot. When the pressing is finished, the oil collected in the clarification drum is skimmed off. The mass to be clarified is boiled, in the traditional manner, to obtain as much oil as possible. Finally, all oil is dried by heating and poured into clean drums for storage.

Results and experience

It appeared that about 10 women were required to process 500 kg of bunches into about 90 kg (or 96 l) palm oil in an 8 hour working day. The oil is sold on the open market. In fact more women assisted in the processing of this quantity. However, some did really work hard, while others were only looking for light jobs. To organize the processing and establish a good co-operation within the group, leadership is required.

Technically, no special problems arose. However, the understanding of the process gave some difficulties. A refresher session appeared to be necessary.

The operation itself is an example of a unit acting as an enterprise. In principle, such a set-up requires quite good management abilities, such as procurement of rawmaterial, sale of products as well as the operation of the unit. To simplify financial management, a service mill might be much more interesting. As only the money for the maintenance and amortization of the equipment and the labour has to be catered for, such an arrangement might be more easily realized.

What remains crucial in the case of a service mill is the leadership to maintain a good working spirit and discipline. The group could contract the processing of a certain quantity of bunches of palmfruit, against a payment in cash or in kind.

Finance

The bunches were made available to the group at F CFA 20 per kg. The oil could be sold at F CFA 200 per l. Since there are seasonal price fluctuations, it would seem to be more beneficial to store the oil for later sale. However, this would require much more working capital. With a reasonable profit to be made by selling the oil directly, such a venture is not advisable. It complicates financial management unnecessarily. Total required investment in equipment is about F CFA 800 000.

Alternative possibilities

As an alternative to the KIT process, using a hand press, at least partly motorized alternatives are a possibility. There is the TCC pounding machine to replace the handpounding and the CALTECH or COLIN type press (SPEICHIM M-10) to replace both pounding by hand and pressing. As described under 2.2.1, 8 women can process about 600 kg of palmfruit per day using the TCC system. This would mean that about 900 kg bunches would have to be cooked and threshed. For the threshing alone, 4 women extra would be required. The motorized version of the CALTECH can process 200 kg steamed fruit per hour. For this only four persons would be required. However threshing would require additional manpower.

For an economic discussion, the threshing problem is better omitted. Chapter 4 will therefore compare the economic performance of the KIT system (8 women and about 400 kg palmfruit per day, giving 112 kg or 120 l palmoil) with alternatives such as the TCC system (8 women and about 600 kg palmfruit per day, giving 148 kg or 159 l palmoil) and the CALTECH (4 persons and about 800 kg palmfruit per day, giving 211 kg or 227 l palmoil).

4. Financial Analysis of the Case Studies

The case studies presented in Chapter 3 include different improved processes for extracting shea nut butter and oil from sunflower seed and oil palm fruit as applied in current projects in Mali, Zambia and Togo. In the present chapter, these projects are financially analyzed in their specific local contexts, which means that all costs and prices involved were taken from recent field experience. It also means, that the results presented are not necessarily the same for other countries or for other times in the future; since, for example, margins between rawmaterial and sales prices, wages and other variables might change considerably.

For each of the case studies, alternatives were calculated, using the same costs for variables like local wages, sales price of oil and by-products, etc. In this way, the profitability of the processes described as case studies can be compared with the technical alternatives.

Finally, for all the case studies and their alternatives a sensitivity analysis has been made, the results of which indicate the profitability of each process in relation to certain critical assumptions. Keeping all others constant as originally, the variables that were modified for this analysis are in each case:
– shea butter/oil recovery of the process,
– capacity utilization of the equipment,
– local wages.

The results are illustrated in graphical form (see Figures 22, 23 and 24). As a guide to interpreting these figures, one might say that the steeper the curve, the more sensitive is the process to a modification of a specific variable.

For all calculations, the same procedure was applied and the same indicator used. The procedure is described in the „Manual for the Preparation of Industrial Feasibility Studies" by the United Nations Industrial Development Organization (UNIDO). Following this procedure, the input-data for all the cases was entered into the computer programme COMFAR (Computer Model for Feasibility Analysis and Reporting), which computes a whole set of output tables. Computations were done for a 10 year project period in each case. For reasons of space and simplicity, the output tables have been reduced to only one indicator.

As the indicator for the results of all case studies, the so-called „internal rate of return" (IRR) was chosen, which is a percentage figure and can – in simple terms – be compared with the long-term interest for a bank deposit. For any investor (i.e. in equipment for oil processing), the IRR is an indicator of whether it is more profitable to put his money in a bank account or to invest in this or that kind of technology and start production under given assumptions. An explanation on how to calculate the IRR is given as Annex 3.

One might argue that the procedure and the indicator are neither appropriate nor relevant for a traditional context where usually no money value is attached to inputs like labour, energy, etc. Nevertheless, a proper financial calculation even for the traditional technologies can introduce a scale on which all alternatives can be measured. In this way, the IRR can serve as a uniformly comparable figure which allows the analysis of profitability. However, it must be kept in mind that it is a capital-based indicator.

4.1 Shea nut processing in Mali

In Mali, the Franc CFA is the accounting currency with F CFA 50 equal to 1 French Franc. As the discounting rate (long-term bank deposits), 10 % was assumed.

Table 13 provides an overview of the costs and results of the improved process compared to the traditional process for

Table 13: **Assumptions for Shea Nut Processing in Mali**
(per year in F CFA, unless stated otherwise)

Input Data	KIT Process	Traditional Process
initial investments		
– land	no costs	no costs
– building	100 000	no costs
– machinery/equipment	450 000	14 000
current investments		
– every year:		
traditional tools	1 000	500
– after 5 years:	hydr. jack	
	100 000	
production costs		
(at maximum capacity = 300 days per year, 8 hours per day)		
– Shea Kernels	15 000 kg	6 000 kg
price per kg	30	30
– utilities	hydr. oil	
	10 000	
– fuel wood	18 750	30 000
– wages		
(at 25 F CFA/h)	300 000	240 000
– maintenance	10 000	
– spare parts	35 000	
– marketing	no costs	no costs
production programme and sales		
– shea butter	962.5 kg	50 kg
price per kg	300	300
working capital requirements		
– for rawmaterial	no costs	no costs
– for utilities	360 days	no costs
– for spare parts	360 days	no costs
source of finance		
– local equity	275 000	14 000
– subsidies	275 000	—

shea butter extraction. The assumptions for the improved (KIT) process are based on the experiences gained so far with the GTZ/GATE/DMA project. The figures are based on field experience in Mali and can be interpreted in the following way:

For the GTZ/GATE project, *land* is made available free of costs by the village and – as with the traditional process – can not be accounted for. The *building* needed is a small traditional one with bricks and a roof of corrugated iron. It is to last for the whole project period but has no value afterwards.

The *machinery* for the KIT process, which for all major parts is expected to last for 10 years, consists of a hydraulic press, an expel stand, oven parts, a few buckets, mortar and pestles. The *equipment* for the traditional process includes mortar and pestles, basins, calabashes, crushing stones and other traditional material.

As *current investments*, the traditional equipment has to be renewed every year. For the KIT machinery, a new hydraulic jack has to be accounted for after 5 years.

Production costs are first based on the maximum capacity of the processes, assuming 300 working days per year. For both processes, shea nuts are the only rawmaterial, costing F CFA 30 per kg on average. The KIT process needs hydraulic oil (10 liters per year at F CFA 1000 per l), an annual technical check-up of the jack and some spare parts (rubber seals, a pump valve). Since fuel wood has little commercial value in rural areas, its price was calculated as the working time needed to collect it (for the KIT process: 2.5 working hours per production day; for the traditional process: 4 working hours per production day). The wages applied are the same as for the actual oil processing, in which 5 women participate with the KIT process and 4 in the traditional one. Marketing costs have not been calculated, since the weekly visit to the market is a social event in any case.

The *production programme and sales* give an indication of the use of the equipment as a percentage of its maximum capacity. The actual yearly production, however, is determined by the availability of rawmaterial, which can – in the case of wildgrowing shea nut trees – be zero in one year and high in the next.

For the KIT process, it was assumed that 50 women form a group, in which each woman has on average 50 kg of shea kernels collected. Processing 50 kg of kernels per day, this raw-material supply would last for 50 working days or 16.6 % utilization of maximum capacity. Assuming an average shea butter recovery of 38.5 %, the process would give 962.5 kg of shea butter per year.

For the traditional process, the 4 women in the group provide together 200 kg of kernels to be processed, which would last for approximately 10 (8 hour) working days or the equivalent to 20 days with 4 actual working hours. The percentage of shea butter recovery was assumed to be 25 % on average, giving 50 kg of shea butter per year. In both cases, the price of the shea butter is F CFA 300.

For the *working capital requirements*, it has to be considered that shea nuts are stored traditionally by rural women as a kind of „bank account" for which, however, no interest can be calculated. For the KIT process, hydraulic oil and spare parts should be stored for a full production year.

As far as the *source of finance* is concerned, it has to be mentioned again that rural women not only have very little capital to invest, but also have difficulties in getting a credit. In the GTZ/GATE project, the total initial investment (building, machinery and equipment) is therefore subsidized with 50 % of its cost (= F CFA 275 000). For the feasibility calculation, however, the subsidy is normally treated as part of the equity capital and has no influence on the profitability of the project.

On the basis of the assumptions described above, which are termed „standard cases" below, the results for both processes are summarized by the calculated internal rates of return (IRR). These are for:
– KIT process 21.91 %,
– traditional process – 8.97 %.

These results mean that the traditional process, quite clearly, is not viable in financial terms, if the critical variables (shea butter recovery rate and wages for processing) are realized as assumed. In practical terms, these results mean that the women involved in the traditional process are not able to realize the assumed wages, i.e. they work in fact for less money per hour (compare sensitivity analysis below). However, since there are very few employment opportunities for women in rural areas of Mali, even this low income is accepted.

The „standard case" of the KIT process, on the other side, produces a positive IRR of roughly 22 %. This means that the financial return, which can be expected from an investment in this process, is considerably better than a bank deposit (assumed as 10 %). The figure also means that the expected return is quite acceptable compared to what is normally expected from an investment into manufacturing industries in developing countries. Depending on the risk he takes (especially on the marketing side), a private investor would – as a rule – feel fairly safe, if his investment can be expected to produce about double the discounting rate (in this case 20 %). Since marketing of shea butter appears to pose no problems in Mali, the calculated IRR of about 22 % can be seen as well promising.

To illustrate the practical implications of the subsidy in the case of the GTZ/GATE project, it is possible (but not sound in the financial analysis of the process) to calculate the IRR only on the basis of the equity provided by the village: Subtracting the subsidy from the total investment, but still having the same cash inflow from sales, the IRR would be more than 50 %. This figure, of course, would indicate a very attractive return for the women concerned.

As shown in the summary sheet for shea nut processing in Mali (Figure 22), variations of the critical variables produce quite different results for the profitability of both processes:

Figure 22, A, indicates the expected IRR in relation to the rate of shea butter recovery. For the standard case of the traditional process, an average recovery rate of 25 % was assumed, which results in a negative IRR. Assuming 30 % recovery for the same process, the IRR would be just over 13 % (i.e. positive). A recovery rate of about 27 % would make the IRR zero; i.e. the process would at least not produce financial losses (i.e. a negative, discounted, cummulated net cash flow).

For the KIT process, the same value (IRR = 0) starts with a recovery rate of about 25 %. Since the investment in this case is relatively big, however, so-called opportunity costs should be considered (at a

Figure 22: Summary Sheet, Shea Nut Processing in Mali. Graphs are approximations.

A) IRR in relation to shea butter recovery
   (capacity utilization and wages as in standard case)

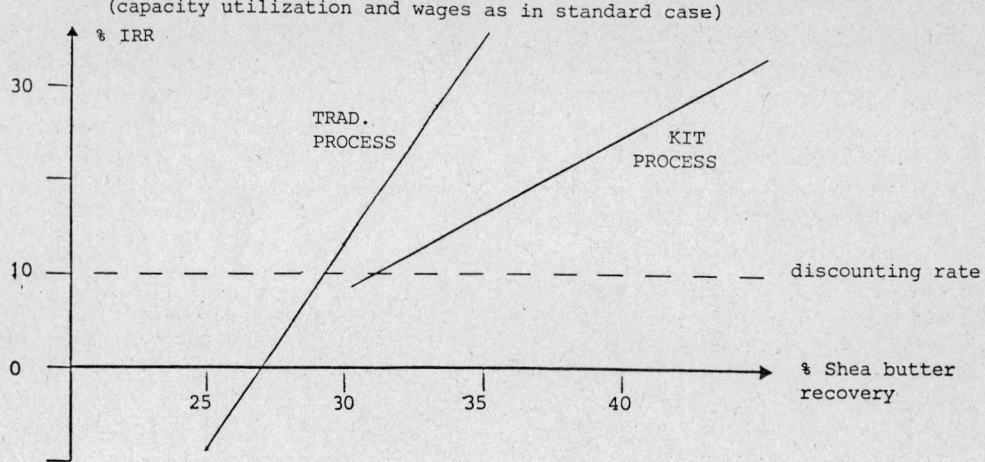

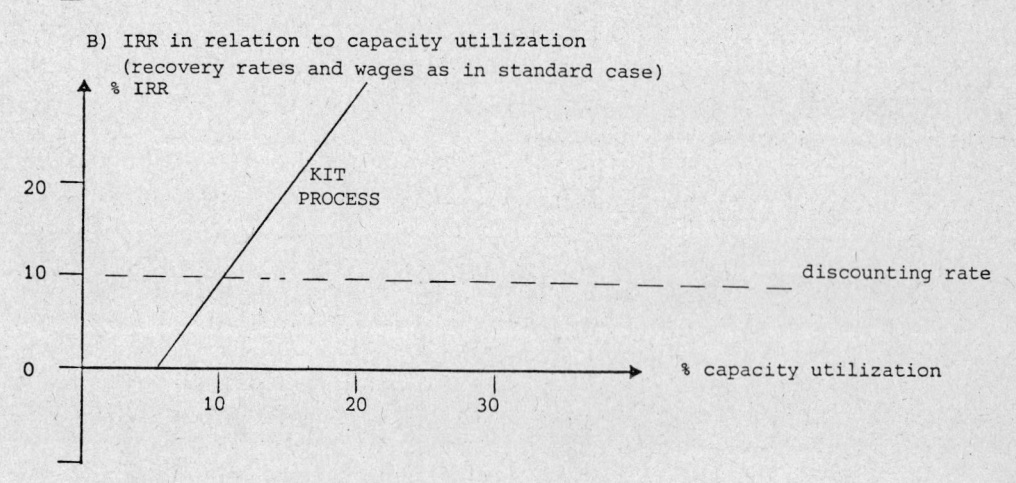

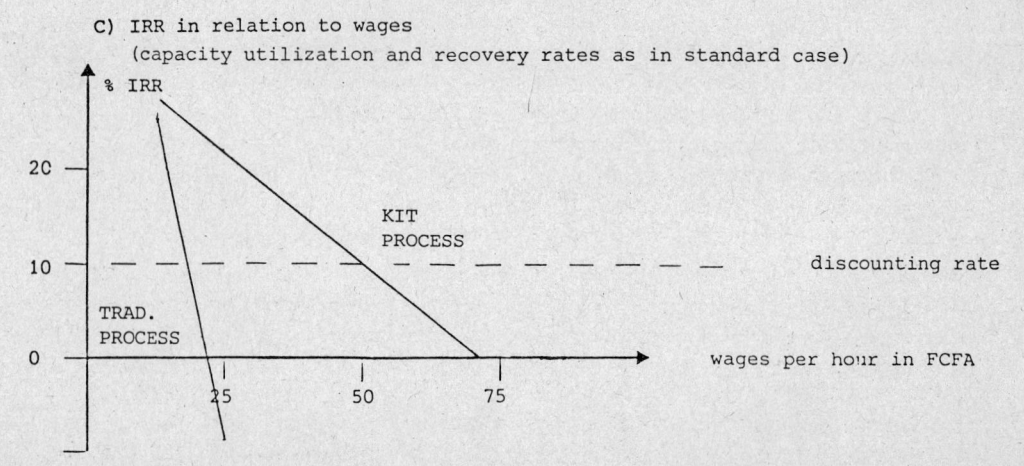

B) IRR in relation to capacity utilization
   (recovery rates and wages as in standard case)

C) IRR in relation to wages
   (capacity utilization and recovery rates as in standard case)

discounting rate or IRR = 10). The KIT process, then, becomes financially interesting, if recovery rates of 32 % and more can be realized. At a recovery rate of 42 %, which is reportedly about the best possible, the IRR would be about 28 %.

Figure 22, B, indicates the expected IRR in relation to capacity utilization (percentage of maximum capacity at 8 working hours per day and 300 working days per year). As mentioned earlier, the bottleneck factor for the utilization of the equipment might be the availability of the rawmaterial, i.e. how many shea nuts the women can collect on average. For both processes, 50 kg of kernels per woman per year had been assumed as the standard case. However, the collected amount of nuts is reported to be higher in many cases. On the other hand, less raw material might be available in years when shea nut trees bear only little fruit. A long-term deterioration of this situation due to a shortage of shea nuts (possibly in consequence of ecological problems such as desertification) has not been considered for the analysis, but might be of importance in the future.

For the traditional process, this variation has (almost) no influence on the IRR, because the investment for equipment, etc. is minimal. For the KIT process, however, an increased shea kernel supply of 60 kg per woman per year (equivalent to 20 % capacity utilization) would bring the IRR to about 30 %; a reduced shea kernel supply of 40 kg (13 % capacity utilization) would reduce the IRR to about 15 %. For the 50 women in the GTZ/GATE project, the KIT process requires a minimum of about 14 kg shea kernels per woman per year to result in an IRR of 0.

Figure 22, C, indicates the expected IRR in relation to wages calculated for collecting firewood and for the actual shea nut processing. In the financial analysis of both processes, labour is the most important production cost factor. Seen from the perspective of the concerned women, the money realized from selling shea butter (wages) is a direct indicator of their available cash income and/or an indirect indicator for the additional spare time.

As can be seen from the graphs, both processes are highly sensitive to labour costs. Most interesting seems the point at which the IRR becomes 0; i.e. the maximum wages that can be realized without running into financial losses. For the traditional process, this point is reached at about F CFA 22; for the KIT process, this point is reached at about F CFA 70 (or F CFA 50 considering the opportunity costs of capital).

For the women participating in the GTZ/GATE project, this result means that their available cash income is about three and a half times the previous (traditional) level without increasing their time spent on processing the nuts. Since the process is limited mostly by the available rawmaterial, the final result for the women could also mean that they can save up to 70 % of the time previously spent on shea nut processing. To give an example: To earn F CFA 1000.– the women in the traditional process have to work for about 45 hours; with the KIT process, it takes only about 13 working hours.

4.2 Sunflower seed processing in Zambia

In Zambia, the Kwacha (K) is the accounting currency. Although the Kwacha is currently under strong pressure, the exchange rate was taken as 8 K for the US $ to calculate the investment in machine-

Table 14: **Assumptions for Sunflower Seed Processing in Zambia**
(per year in Kwacha, unless stated otherwise)

Input Data	Hand-operated Equipment	Simon-Rosedowns	Reinartz
initial investments			
– land	←——————— no costs ———————→		
– building	10 000	10 000	10 000
– machinery	20 000	81 000	180 000
– equipment	1 000	1 200	2 000
– seedstock up	12 500	21 000	42 000
current investments			
– first year: further seed stock up	12 500	21 000	42 000
– every year: equipment renewed	300	600	1 000
– after 5 years:	gear box/ decorticator 2 000	diesel engine 15 000	diesel engine 25 000
production costs (at maximum capacity = 250 days per year, 8 hours per day)			
– sunflower seeds (kg)	37 500	62 500	125 000
price per 100 kg	84	84	84
– utilities	no costs	diesel 4 450	diesel 15 500
– fuel wood	675	no costs	no costs
– wages	(4.5 K/d) 6 750	(10 K/d) 5 000	(10 K/d) 10 000
– maintenance	675	1 500	3 000
– spare parts	300	10 000	15 000
– marketing	675	1 000	2 000
production pogramme and sales			
– sunflower oil (litres), year 1	3 000	5 000	14 000
year 2 onwards	6 000	10 000	28 000
price per litre	7	7	7
– oilcake (kg), year 1	6 600	19 000	33 000
year 2 onwards	13 200	38 000	66 000
price per 100 kg	50	30	30
working capital requirements			
– for rawmaterial	←——————— seed stocks as above ———————→		
– for utilities	no costs	30 days	30 days
– for spare parts	360 days	360 days	360 days
– for finished products	7 days	7 days	7 days
source of finance			
– local equity	36 200	53 200	100 000
– local loan	20 000	81 000	176 000

ry. As the discounting rate (long-term bank deposits), 10 % was assumed. For a loan from a local bank (6 years, 1 year grace period), an interest rate of 20 % was seen as realistic.

Table 14 is to give an overview on costs and results for the use of the hand-operated equipment developed by KIT, and as alternatives the expellers from Simon-Rosedowns and Reinartz. The figures are based on field experience in Zambia and can be interpreted in the following way:

As in Mali, *land* in rural areas in Zambia can be assumed to be provided free of costs by a group or the village. The *building* would in all cases be of rather small size and is assumed to have a residual value of 50 % after the project period.

The *machinery* for the hand-operated process consists of a decorticator, a winnower, a roller mill, a heating oven, a spindle press and other equipment like weighing scales, buckets, etc. The two technical alternatives consist both of diesel-powered expellers, the first from Simon- Rosedowns (MINI 40), the second from Reinartz (AP VII). Both expellers (not the diesel engines) are to last for ten years and the delivery prices include spare parts for two years.

Since seasonal price fluctuations for sunflower seeds are considerable, a one-year *stock* of rawmaterial has to be bought at low prices just after the harvest before the first production year. A second seed stock-up of the same amount is due as a current investment within the first production year to be able to double production from the second year onwards (see production programme below).

Further *current investments* refer to some equipment to be renewed every year (cleaning sieve, scales, buckets, drums, etc.), and machinery parts after 5 years comprising a new gear box for the decorticator and, for the two technical alternatives, new diesel engines.

Production costs involve the annual expenses for the raw- material seeds (except in the 10th year of production). For the two expellers, utilities (diesel, greasing oil, etc.) have to be calculated; for the hand-operated equipment, fuel wood (as valued by the wages for collecting) applies. Wages for the hand-operated process can be calculated lower (6 men with 4.5 K per day) than those for the alternatives (2 men and 4 men respectively with 10 K per day), since the equipment demands at least semi-skilled labour. In all cases, costs for maintenance and spare parts were calculated; marketing costs refer to the common practice of door-to-door sales or running a small shop.

The *production programme* in all three cases is geared to 40 % of maximum capacity for the first year of production, to be doubled in the second year and afterwards kept on this level until the end of the project period. Some caution is, however, advisable in trying to apply this assumption to the real Zambian context, since no thorough market investigation could be done for this publication to verify whether the whole production (especially of the Reinartz version) could be easily sold. In all cases, the oil-cake as a by-product has a significant impact on the feasibility of the projects.

Apart from the stock-up on rawmaterial, working capital is required for utilities and spare parts (bearings, scrapers, leather belts for the first case; choke rings, rings, screws for the second; and scrapers, parts of the screw, cage-bars, etc. for the third). The finished products, which are usually sold only once every two weeks, have to be stored for 7 days on average.

Figure 23: Summary Sheet, Sunflower Seed Processing in Zambia. Graphs are approximations.

A) IRR in relation to oil recovery
   (capacity utilization and wages as in standard case)

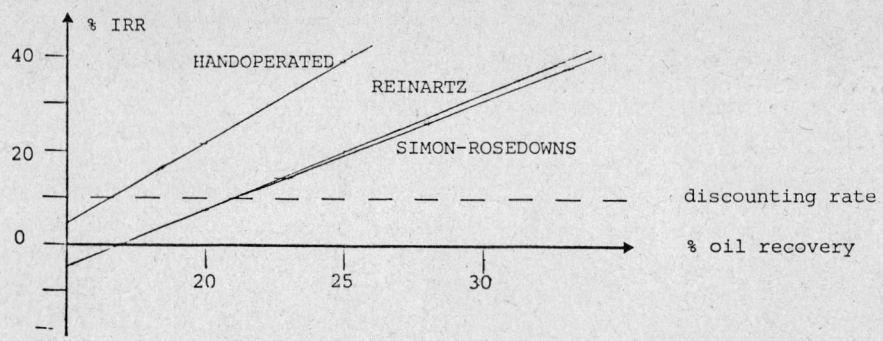

B) IRR in relation to capacity utilization
   (recovery rates and wages as in standard case)

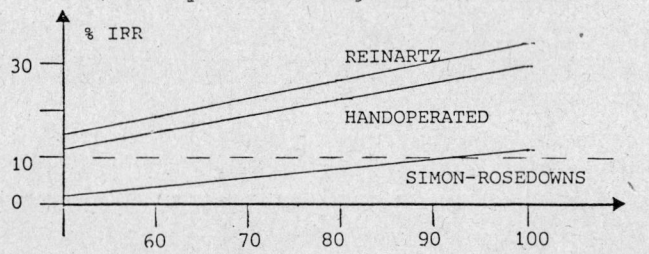

C) IRR in relation to wages
   (capacity utilization and recovery rates as in standard case)

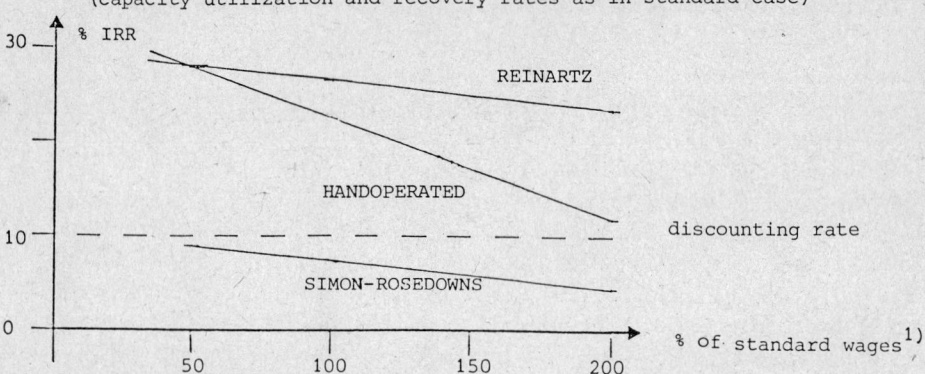

[1] Assumed standard wages per day:
- 100 % = 4,5 K for handoperated process (unskilled labour)
- 100 % = 10 K for expeller (semi-skilled)

In all cases, a local loan will probably be necessary in addition to equity capital to cover the investments. As can be seen from the figures in the above table, the total volume of the investment, at least for the Reinartz expeller, is not the smallest scale any more and could probably only be managed by a single investor or a well organized formal cooperative.

The profitability of the three versions of equipment for sunflower seed processing is again given by the internal rate of return (IRR) as the major indicator for the *financial evaluation*. The following figures were calculated:
– hand-operated equipment 22.47 %,
– Simon-Rosedown expeller 7.54 %,
– Reinartz expeller 26.55 %.

Under the assumptions described above (standard cases) for Zambia, the Simon-Rosedowns expeller cannot be seen as a feasible enterprise, partly due to expensive spare parts. The hand-operated equipment undoubtedly has its strength for small-scale processing with little capital to invest and to risk. The Reinartz expeller requires a sizable market outlet, considerable volumes of rawmaterial supply and investment and management skills in general, but has its advantages due to a relatively good rate of oil recovery.

As for Mali, the standard cases in Zambia have been further analyzed by modifying some critical input data. The results of this *sensitivity analysis* are illustrated in Figure 23 and can be summarized as follows:

Figure 23, A, indicates the expected IRR in relation to the rates of sunflower oil and oilcake recovery. (Recovery rates of oil and cake are seen as complementary; i.e. if in the modifications the oil recovery rate was reduced by 5 %, the oilcake recovery would be increased by 5 % and vice versa. The difference is, of course, the sales price of the two products.) Depending on the quality (oil content) of the seeds and the condition of the equipment, the oil recovery varies over a considerable range. As illustrated by the graphs, all three processes are financially very sensitive to varying recovery rates.

For the hand-operated equipment, an oil recovery rate of 20 % was assumed as the standard case, for which about 22 % IRR can be expected. The same process would produce more than 39 % financial return with 25 % oil recovery; below about 13 % oil recovery, the project would run into financial losses.

The Simon-Rosedowns equipment, which is just barely feasible to operate with standard case assumptions (20 % oil recovery) would make a financial loss with oil recovery rates below 17 %. The process, however, gives good results (IRR more than 20 %), if the recovery rate can be brought up to 25 %.

The process with the Reinartz expeller shows an IRR = 0 point at a minimum recovery rate of about 15 %. Within a more realistic range (23 % to 33 % oil recovery, taking 28 % as the standard case) the process gives very attractive financial returns of between 14 % and 37 % (27 % for the standard case).

Figure 23, B, indicates that all three processes are less sensitive to varying rates of capacity utilization which can be explained by the high proportion of variable costs in the production. As for the standard cases, all the variations illustrated by the graphs assume that for the first production year actual production is only half of what is produced from the second year onwards.

At full production (100 % capacity utilization from 2nd year), the process with the

Reinartz expeller gives the best financial returns (34 %), followed closely by the hand-operated process (29 %). The Simon-Rosedowns equipment performs still rather poorly (IRR = 11.5 %) under these optimal conditions.

Figure 23, C, shows the relation of the IRR to the wages calculated for collecting fuel wood and the actual processing. The hand-operated equipment requires, of course, the most labour-intensive process and is therefore the most sensitive one in terms of labour costs (or wages realized). In this process, maximum wages of about 13.5 Kwacha per day could be paid before the project would run into financial losses. For the process using the Simon-Rosedowns expeller, this point would be reached at around 14.5 Kwacha per day. The Reinartz process is so capital intensive that wages play only a minor role for its profitability.

4.3 Oil palm fruit processing in Togo

As in Mali, the Franc CFA is the accounting currency in Togo. A 10 % discounting rate and loan conditions of 18 % (6 years, 1 year grace period) were assumed.

Table 15 gives an overview on costs and results for the use of the hand-operated equipment developed by KIT, and as alternatives the process developed by TCC, Ghana, and the CALTECH expeller. The figures are based on field experience in Togo and can be interpreted in the following way:

As for the other case studies, *land* in rural areas in Togo is assumed to be available free of costs. The *building* in all three alternatives consists of a simple, large (60 m²) shed, which would have no residual value after 10 years.

The *machinery* for the hand-operated process consists of a UNATA spindle press, cooking kettles, buckets and other equipment like oil containers and pounding sticks, which are to be renewed every year. For the TCC process, one TCC pounding machine, two TCC screw presses, one large cooking kettle and some buckets are required. The engine of the pounding machine is to be renewed after 5 years. The CALTECH process only includes as expenses for machinery the expeller and some drums. The expeller is driven by a petrol engine, which probably has to be renewed in the fourth and seventh year of production. A new screw for the expeller should be accounted for after 5 years.

Production costs at maximum capacity (250 working days per year) would involve loose oil palm fruits as the only raw- material in the above given quantities and for a price of F CFA 35 per kg. For the motor-driven alternatives, costs for utilities are for diesel/petrol and lubricants. The costs for fuel wood are again calculated on the basis of the wages for collecting. Wages for the hand-operated and the TCC process cover 8 women at F CFA 400 per day; with the CALTECH version, 2 women at F CFA 400 per day and 2 semi-skilled labour (i.e. usually men) at F CFA 800 per day get employed. Whereas maintenance is mainly for the engines and minor repairs of the equipment, spare parts include drums, kettles, oil filters, etc. In all three cases, costs for administration and marketing have been accounted as one working day once a fortnight.

The *production programme and sales* are, in all three standard cases, based on the assumption that the maximum capacity is only used to 10 % in the first year (= 25 working days) and to 20 % from the second year onwards (= 50 working days per year). The reason for this limitation

Table 15: **Assumption for Oil Palm Fruit Processing in Togo**
(per year in F CFA, unless stated otherwise)

Input Data	Hand-operated Equipment	TCC-Process	CALTECH Expeller
initial investments			
– land	no costs	no costs	no costs
– building	200 000	200 000	200 000
– machinery	550 000	1 300 000	2 550 000
– equipment	10 000	15 000	20 000
current investments			
– every year: equipment renewed	10 000		
– year 4&7:			petrol engine 100 000
– after 5 years:		diesel engine 500 000	expeller screw 10 000
production costs (at maximum capacity = 250 days per year, 8 hours per day)			
– oil palm fruit (kg)	100 000	150 000	200 000
price per kg	35	35	35
– utilities	no costs	diesel 260 000	petrol 105 000
– fuel wood	100 000	100 000	150 000
– wages 400/800 F CFA/day	800 000	800 000	600 000
– maintenance	10 000	20 000	40 000
– spare parts	20 000	30 000	30 000
– administration	25 000	25 000	25 000
– marketing	25 000	25 000	25 000
production programme and sales			
– palm oil (litres), year 1	3 000	3 975	5 675
year 2 onwards	6 000	7 950	11 350
price per litre	200	200	200
– palm kernels (kg) year 1	700	925	1 325
year 2 onwards	1 400	1 850	2 650
price per kg	60	60	60
working capital requirements			
– for utilities	no costs	30 days	30 days
– for spare parts	360 days	360 days	360 days
– for finished products	30 days	30 days	30 days
source of finance			
– local equity	260 000	515 000	1 000 000
– local loan	500 000	1 000 000	1 770 000

Figure 24: Summary Sheet, Oil Palm Fruit Processing in Togo. Graphs are approximations.

A) IRR in relation to oil recovery
(capacity utilization and wages as in standard case)

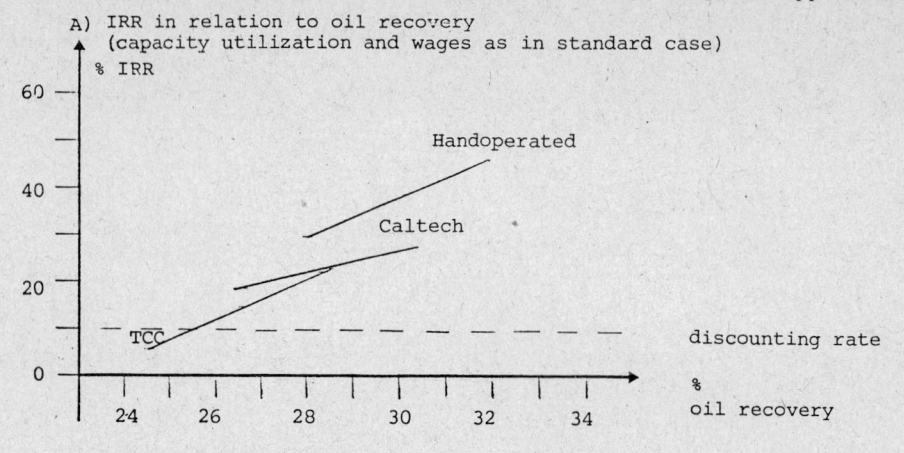

B) IRR in relation to capacity utilization
(recovery rates and wages as in standard case)

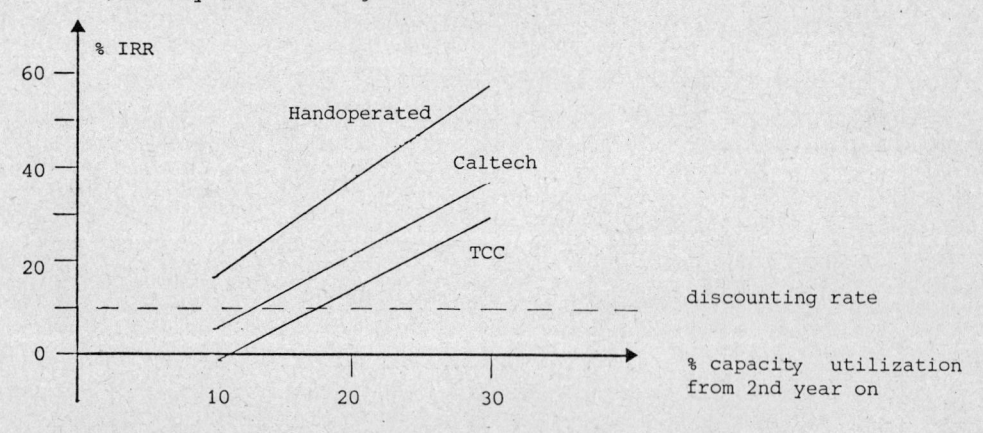

C) IRR in relation to wages
(capacity utilization and recovery rates as in standard case)

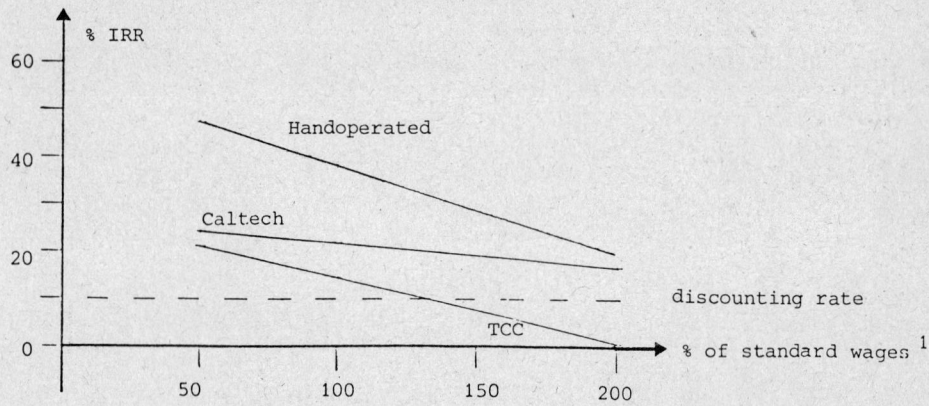

[1] Assumed standard wages per day are 400 F CFA for women and 800 F CFA for men.

is, in the local context, the availability of rawmaterial. Palm oil and palm kernels are sold at prices fixed by the Government. Palm kernels are not processed further because this involves either a very labour or very capital intensive procedure which is financially not attractive in comparison to the sales price of the kernels.

As *working capital,* spare parts and a reserve of petrol/diesel fuel is required for the motor-driven alternatives. Finished products are usually taken on stock for a month.

As *source of finance,* a loan from a local bank is probably required in addition to equity capital to cover the investments.

For the the *financial evaluation* of the three versions of equipment for oil palm fruit processing, again, the internal rate of return (IRR) has been chosen as the major indicator. The following figures were calculated:

– hand-operated equipment 38.35 %
(30.0 % oil recovery),
– TCC process 14.91 %
(26.5 % oil recovery),
– CALTECH expeller 22.00 %
(28.4 % oil recovery).

Under the assumptions made for the standard cases in Togo (for oil recovery rates, see figures in brackets above), all three processes would give satisfactory returns on invested capital, the hand-operated equipment even producing a very good rate of return. The *sensitivity analysis* which is illustrated in Figure 24, however, indicates that the financial returns for all three processes react rather sensitively to changing oil recovery rates and wages and very sensitively to changing capacity utilization.

Figure 24, A, indicates that for the hand-operated equipment the expected IRR would still be very satisfactory (just below 30 %), if the the oil recovery rate were 2 % lower than the assumed standard case. The same reduction in oil recovery for the TCC equipment would, on the other side, produce an IRR of less than 6 %, which is less than the interest on a bank deposit and therefore not an attractive investment.

Figure 24, B, indicates that capacity utilization (i.e. rawmaterial supply) is a very critical factor for all three processes. For the standard cases, 50 working days per year (or 20 % utilization of maximum capacity) had been assumed. If this is reduced to 25 working days, the TCC process would already run into financial losses. At 75 working days per year, on the other side, the TCC and the CALTECH equipment would show very good results (IRR = 27 % and 35 % respectively.); the hand-operated equipment would even allow financial returns of more than 56 %.

Figure 24, C, indicates that the hand-operated process is, naturally, most sensitive to changing labour costs, but could financially still tolerate maximum wages (unskilled workers) of about F CFA 1,200 per day. A project using the TCC equipment, on the other side, would go bankrupt, if wages of more than F CFA 800 per day were paid. The CALTECH equipment is less sensitive in this respect.

5. Selected Equipment

5.1 Hand-operated equipment

5.1.1 Hand-operated processing of palm fruit

Local manufacturing:

ENDA, Dakar, Senegal
ENDA promotes the use of a locally manufactured press. This press is used to reprocess the fibre that remains after the traditional process to obtain the residual oil. In this way the oil recovery is improved by 40 %.

The same result can be obtained by using this press for the semi-traditional process (as described in 2.2.1). In this case the fibre is steam-heated before pressing. As the maximum attainable pressure is only 8 bar, the press is not suitable for pressing mixtures of nuts and fibre. (As the dura variety of the Casamance has a very low pulp-to-nut ratio, it is worthwhile applying the KIT process (see 2.2.1). This improves the traditional oil recovery by another 40 %, making the overall oil yield about 80 % of the oil originally present.

The ENDA press consists of a frame built from locally available materials such as wood, iron channel and strip, and equipped with a nut and spindle. In this frame a perforated cage is placed in a container, fitted with an outlet tube, as oil receiver. The press is to be manufactured by the local blacksmith. The spindle is supplied

Figure 25: Oil Press, as disseminated by ENDA.
Source: ENDA, 1981, after Thierno Diedhiou

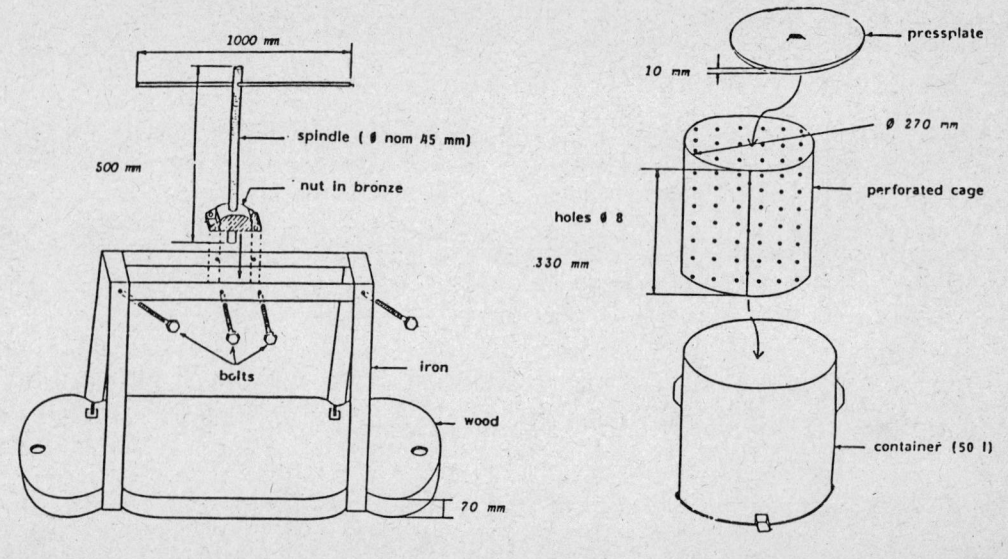

to him by others but the nut is cast from bronze by the blacksmith himself.

The cage content is 19 litres (l). This content is such, that all the fibres of a day's production can be reprocessed in less than an hour in a few pressings.

The involvement of the local blacksmiths has many positive aspects, e.g. ready availability of spare parts and repair service close at hand. The price of the press is F CFA 30 000 (20 000 for materials and 10 000 for the work). The costs of maintenance and repair are also estimated at F CFA 30.000 per year. This includes renewing of the nut, which is subject to wear and needs to be replaced every season.

To improve upon the quality of the press ENDA is designing a reinforced version. Also auxiliary equipment, such as sterilization and heating units, are being designed. A palmnut-cracker, to obtain the palmkernels, is under development as well.

TCC, Kumasi, Ghana

For the pressing of nuts and fibre, TCC developed a sturdy press to be manufactured by local craftsmen from locally available materials. Manufacturing takes place under the guidance of the Suame ITTU, Kumasi.

The TCC Press consists of a table, with a few perforations, 50 cm high. A 60 cm long spindle, welded on a bottom press plate is mounted in the centre of the table. A press cage (about 30 cm diameter and 30 cm high) made from 4 mm thick cylindrical waterpipe and consisting of two halves, is placed on the table around the bottom press plate and with the spindle in the centre. The two halves are hinged at one side and locked with a pin at the other, to allow for easy opening of

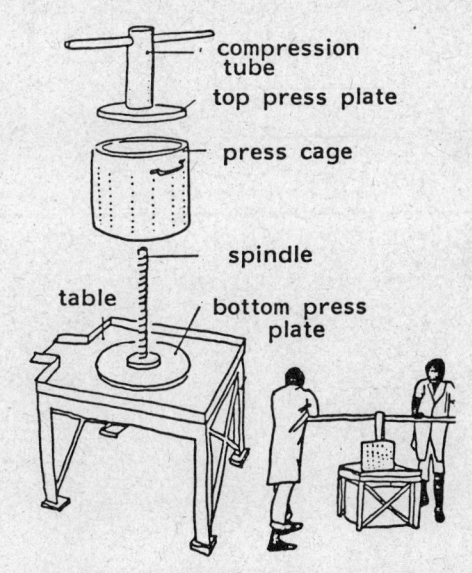

Figure 26: TCC Press. Source: GRET, after Ph. Langley, 1984, p. 22

the cage and removal of the pressed mass. Around the spindle, a top press plate is placed, which is fixed to a threaded compression tube. On the tube a turnbar is welded. The mass is compressed in the cage between the two press plates by turning the turnbar. The turnbar can be equipped with extension tubes to enable women processors to exert enough pressure. The oil is collected below the table and the pressed mass at the side. In 1985 the price of the press amounted to C (Cedis) 29 300 (US $ 560).

The advantages of the TCC press are:
– its sturdiness
– its relatively light construction
– the possibility to exert a maximum pressure of 40 bar on the mass to be pressed, making the press suitable to press mixtures of nuts and fibre.

However the construction of the press, with the spindle in the centre, makes the removal of the press cake difficult and the press unsuitable for the pressing of oil seeds.

Additional equipment, developed by TCC and manufactured under the guidance of the Suame ITTU are:
– a cooking kettle for steaming of loose fruit
– a clarification kettle of the same size as the cooking kettle.

Processing capacity and oil recovery depend on rawmaterial. The performance of a complete TCC unit, which includes a motorized pounder (see Chapter 5.2.1) is given in Chapter 4.3.

Major Exporters:

UNATA, Ramsel, Belgium
UNATA manufactures a sturdy press, developed in cooperation with KIT. The press UNATA 4202 consists of a frame of channel iron in which a cast iron nut is mounted at the top and a flat oil receiver/filling table at the bottom. The press should be mounted on a concrete foundation. A small frame, holding the bolts to fix the press, should be embedded in the concrete.

In the cast iron nut, a steel spindle turns, equipped with a top press plate. A perforated cage (contents 17 l) is filled on the table and placed in the centre of the press just below the top press plate. Pressure can be exerted on the top press plate by rotating the spindle with turnbars that can also be extended with additional bars. The oil flows on the table and is collected at both sides of the table. The press is particularly suitable to press mixtures of nuts and fibre.

The advantages of this press are:
– its sturdiness
– the easy way of entending the turnbars to make it possible to exert a maximum pressure of 40 bar

– the easy way the press cage can be removed, so that pressing can be continued with a second one
– its suitability for pressing oil seeds (with a slightly adapted press cage).

The only part that shows a little wear when the press is used continuously at maximum pressure, is the thrust bearing, where the spindle turns in the top press plate. Frequent greasing is required at this spot.

The price of the press (1987, ex factory) is BF (Belgian Francs) 30 600,-. Local manufacturing is quite possible as this press is presently being manufactured for sunflowerseed processing in Zambia.

Other equipment to be obtained from UNATA includes:
– parts for cooking and reheating drums
– cooking kettles (up to 2 m^3)
– reheating kettle.

Processing capacity and oil recovery depend on rawmaterial. The performance of a complete UNATA/KIT Unit is given in Chapters 3.3 and 4.3.

Usine de Wecker, Wecker, Luxemburg
A hand-operated vertical hydraulic press is manufactured by Usine de Wecker, referred to as type PM 50-83.

The press consists of a frame with a hydraulic ram at the bottom and a press cage hanging on one of the stands of the frame and swinging around it. The cage can be put in three positions: filling, pressing and discharge position.

In the pressing position, the cage hangs above the hydraulic press ram which is equipped with a low and a high pressure hand-operated pump. The filling capacity of the press is 50 l. Its weight is 660 kg.

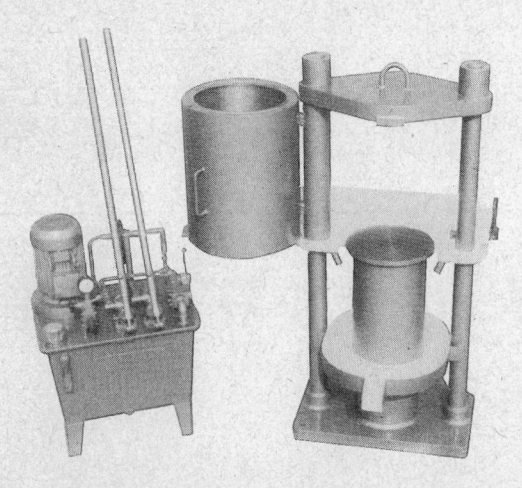

Figure 27: Hydraulic Press, Usine de Wecker, Type 50-83. Source: Usine de Wecker

The price (1987) amounts to DM 14 200 (f.o.b. Antwerp). For the cracking of palm nuts, for obtaining the palm kernels, a hand-operated cracker is available as well.

Processing capacity is in principle three times the UNATA press. However, to attain this capacity the auxiliary equipment, such as cooking kettles, pounder and/or reheating and clarification kettles, should have a comparable capacity. Oil recovery is dependent on the raw-material and which process is followed.

Tool Foundation, Amsterdam, Netherlands
This organization co-ordinates the manufacturing and sale of the 8 L spindle press, designed by KIT. This press is equipped with a steel spindle that turns in a cast-iron nut. The frame is constructed from sheet metal. The weight of this press is only 60 kg. Maximum attainable pressure is 40 kg/cm².

This press is particularly suitable for the so-called „semi-traditional" process for the processing of oil palm fruit. In this

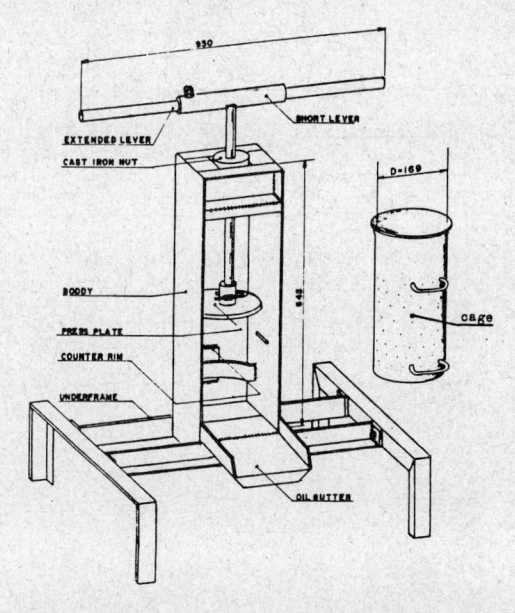

Figure 28: 8 L Spindle Press, KIT/TOOL and Kettles

process the nuts and fibre are separated in water and the fibre is subsequently reheated for pressing. This press can best be used in combination with cooking and reheating kettles made from oildrums. The price of the press is (1986) FL (Dutch Gulden) 1273.

Its capacity is half the UNATA press. Oil recovery depends on the rawmaterial and which process is followed.

5.1.2 Hand-operated processing of oil seeds

Local manufacturing

„Presse Garango", Senegal and Burkina Faso
The „Presse Garango" is a press mainly in use for processing groundnuts. These presses are manufactured by local blacksmiths from locally available materials.

They can be found on markets in Senegal and Burkina Faso. Its construction is, in principle, the same as the press disseminated by ENDA in the Casamance for processing oil palm fruit, which is also based on the „Garango" press. Its use for processing groundnuts compared with the traditional way of preparing oil and „kulikuli" is described by Corbett. As „kulikuli" fetches a good price and the sale of the press cake from the press is uncertain, the traditional process is still financially more attractive.

Project GTZ/GATE, Bamako, Mali
For the processing of shea nuts (Karité), a press based on a hydraulic jack has been developed by KIT. This press consists of a frame in which a table can be moved up by a hydraulic jack. A perforated cage is placed on the table and the mass in the cage is put under pressure against a press block by moving the table upward. The cage content is 8 l and the maximum pressure is 125 kg/m². For expelling the cake from the cage an expel-stand is made as well.

To provide the required pressure a 30 ton lorry jack is applied (make NIKE, from Sweden). Recently it has been decided to equip the jack with an improved pump, which can withstand the forces exerted on it. This pump is based on parts of the 50 ton jack made by NIKE and a specially made pump-cylinder and pump-plunger. To facilitate the cooking process, an oven is required. Press, expel-stand and oven parts are manufactured by a workshop in Bamako, Mali. The price 1987 for a press and expel-stand is F CFA 400 000, for the oven F CFA 40 000. Recently, its manufacture has also been taken up by the IBE (Institut Burkinabé d'Energie), Ouagadougou, Burkina Faso.

Processing capacity and oil recovery are given in Chapters 3.1 and 4.1.

Figure 29: Shea Nut Press, GTZ/GATE.
Source: GTZ/GATE

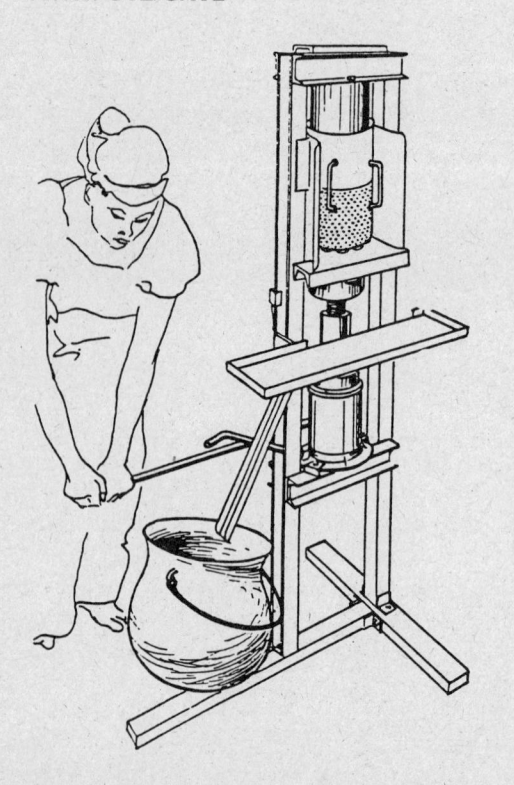

IPI, Dar Es Salaam, Tanzania
IPI has developed a complete line of equipment for the processing of sunflowerseed, according to the dry process. At first only decorticating, heating and pressing at high pressures were applied but with disappointing results. Later a rollermill and an improved „cooking method" were introduced. The results with this equipment seem to be comparable to those obtained with the equipment available from UNATA.

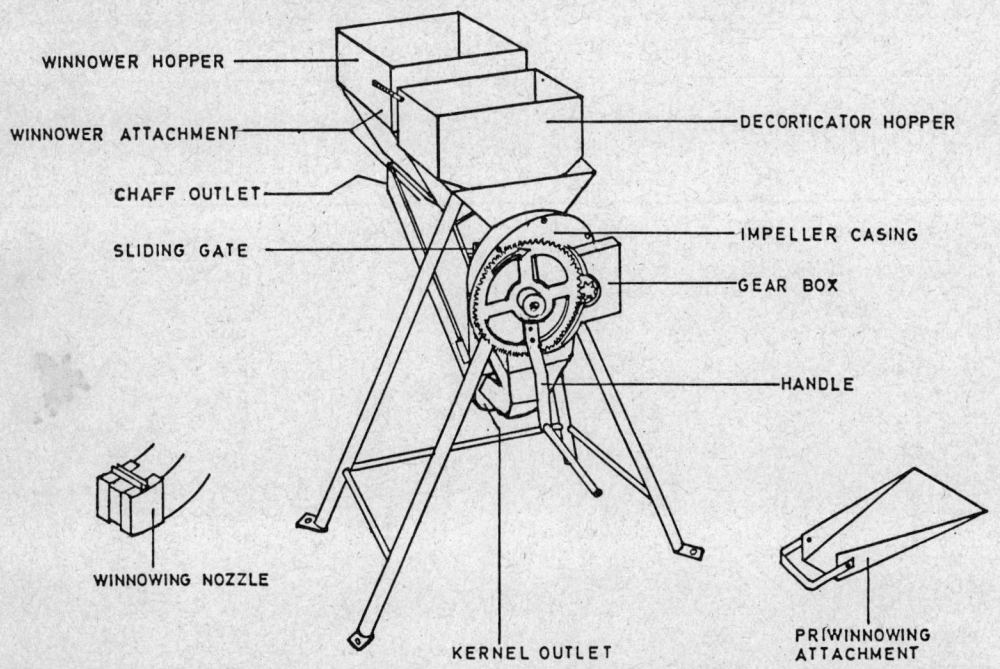

Figure 30: IPI Decorticator/Winnower

Figure 32: IPI Seed Scorcher

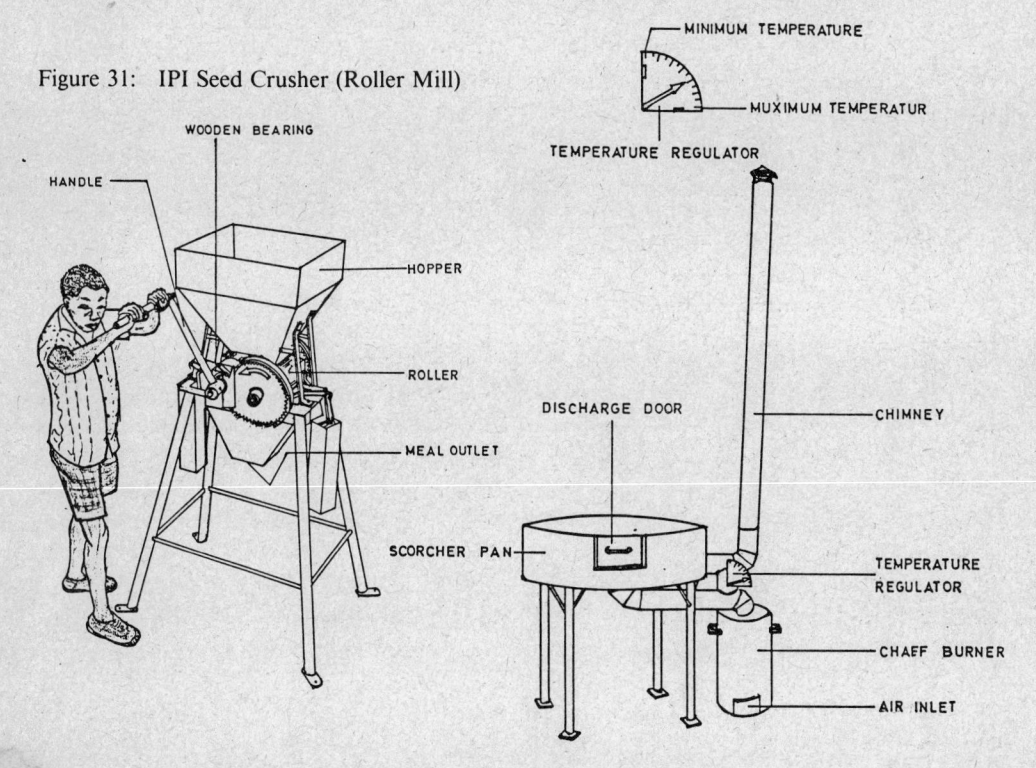

Figure 31: IPI Seed Crusher (Roller Mill)

83

Figure 33: IPI Oil Press

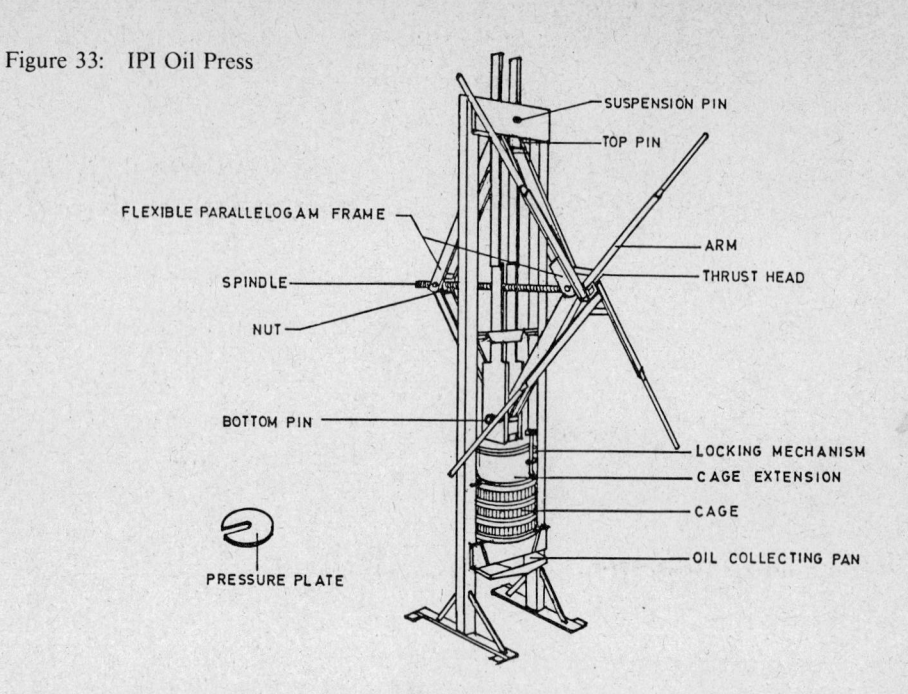

Table 16: **Weight and Capacity of IPI Equipment**

		weight (kg)	capacity (kg/h)
Decorticator/winnower	(Fig. 30)	132	40 (seed)
Seed crusher	(Fig. 31)	118	32 (dec. seed)
Seed scorcher	(Fig. 32)	125	21 (dec. seed)
Oil press	(Fig. 33)	930	21 (dec. seed)

The heart of the equipment, the press, is based on a shear jack system. The press can process 21 kg decorticated, rolled and „cooked" sunflowerseed per batch in a one-hour cycle.

Weight and capacity of the IPI equipment are given in Table 16.

The press is 3.65 m high and requires a sturdy foundation and quite some space. IPI reports (Chungu A.S.) a daily capacity with 7 operators of 210 kg seed (140 kg decorticated seed), giving 42 kg oil. The price of a complete IPI sunflowerseed processing unit is (1987) T. Sh 250 000 (about US $ 4000).

Major exporters

UNATA, Ramsel, Belgium
UNATA manufactures a nearly complete line of equipment for the processing of oil palm fruit and oil seeds. The central piece is the UNATA spindle press, with a cage content of 17 l, and a capacity of about 10 kg per batch in a 30-minute cycle.

Processing capacity and oil recovery are given in Table 9 and for sunflowerseed in detail in Chapters 3.2 and 4.2.

Weight and capacity of the UNATA equipment for sunflowerseed processing are given in Table 17.

Table 17: **Weight and Capacity of UNATA Equipment**

		weight (kg)	capacity[1] (kg/h)
Cracker[2]	(Fig. 14)	85	
Winnower[2]		35	50 (seed)
Rollermill	(Fig. 16)	175[3]	30 (dec. seed)
Heating oven			20–30 (dec. seed)
Oil press	(Fig. 18)	125	20–30 (dec. seed)

[1] with two operators.
[2] in combination.
[3] can be shipped with flywheels to be filled with concrete; in that case the weight is 135 kg.

KIT has reported a daily capacity with 6 operators of 150 kg sunflowerseed (100 kg decorticated seed), giving 29 kg oil and with 15 operators of 300 kg seed (200 kg decorticated seed) giving 58 kg oil.

Prices of these machines are (1987, ex-factory, packed):

– Cracker BF 68 540
– Winnower BF 19 990
– Rollermill BF 57 800
– Oil press BF 30 300

Tool Foundation, Amsterdam, Netherlands
Tool Foundation co-ordinates the manufacture and sale of the sunflowerseed decorticator/palmnut cracker from the Netherlands (see Figure 14). Its price (1987) is about FL (Dutch Gulden) 3000.

5.2 Motorized equipment

5.2.1 Motorized processing of oil palm fruit

Local manufacturing

TCC, Kumasi, Ghana
To replace the hand pounding for digesting the fruit before pressing, TCC developed a motorized pounding machine (see Figure 10). The machine consists of a horizontal cylinder, equipped with an inlet funnel on one side and an outlet on the other. A helical screw rotates within the cylinder driven by an 8 HP diesel engine.

The pounder operates continuously at a rate of about 100 kg per hour. As a TCC press has a capacity of 50 kg per hour, the pounding machine has to be combined with two TCC presses. The pounding time is about 10 minutes (which seems relatively short), since no steam for heating is applied. Oil recovery from tenera fruit is reported to be reasonable: 26.5 l or 25 kg from 100 kg fruit, which is about 17 % on bunches.

In 1985 the price of the pounder amounted to C (Cedis) 28 000 (US $ 535), excluding the diesel engine. Processing capacity and estimated investment for a complete unit is given in Chapter 4.3.

OPC, Douala, Cameroon
OPC manufactures the equipment as developed by APICA. APICA has developed the CALTECH continuous screw press in principle for hand-operation (see Figure 11). However, in practice the machine has to be motor-driven, as the effort required is too much to be contin-

ued longer than 30 minutes per day by the same men. When motor-driven, the press should be equipped with an electric motor or a petrol engine (2.3 HP = 1.7 kW; 2000 rpm).

When motor driven, its capacity is 200 kg cooked tenera palm fruit per hour with an extraction rate of 17 % for bunches. (see Table 8.). The machine is particularly suitable for processing of tenera palm fruit. The result with dura palm fruit is less impressive.

Prices (1987 ex- factory) are:
– Manually operated version (80-105 kg fruit/h): F CFA 1 080 000
– Motorized (petrol) version (150-200 kg fruit/h): F CFA 1 800 000

OPC manufactures also the screw press as originally built by the French firm Colin (now Speichim). This press is available in a manually operated and a motorized version, as well.

Prices (1987 ex-factory) are:
– Manually operated version (110-150 kg fruit/h): F CFA 3 000 000
– Motorized (petrol) version (275-375 kg fruit/h): F CFA 4 110 000

Major exporters

SPEICHIM, Bondy, France
The expeller press, as originally developed by COLIN (see 2.2.1) is at present offered by SPEICHIM. Two sizes are produced, the so-called „single screw press M-10" being the smallest.

This press consists of a perforated cage in which a screw with a special profile turns to realize two functions. After feeding and compacting, the mass is pressed by a counteracting effect of the screw. During the pressing stage, transport of the mass is continued until it reaches the end of the cage. Backpressure is controlled with a control valve at the end of the cage.

Figure 34: Speichim Expeller Press. Source: GRET, 1984, p. 18, after APICA

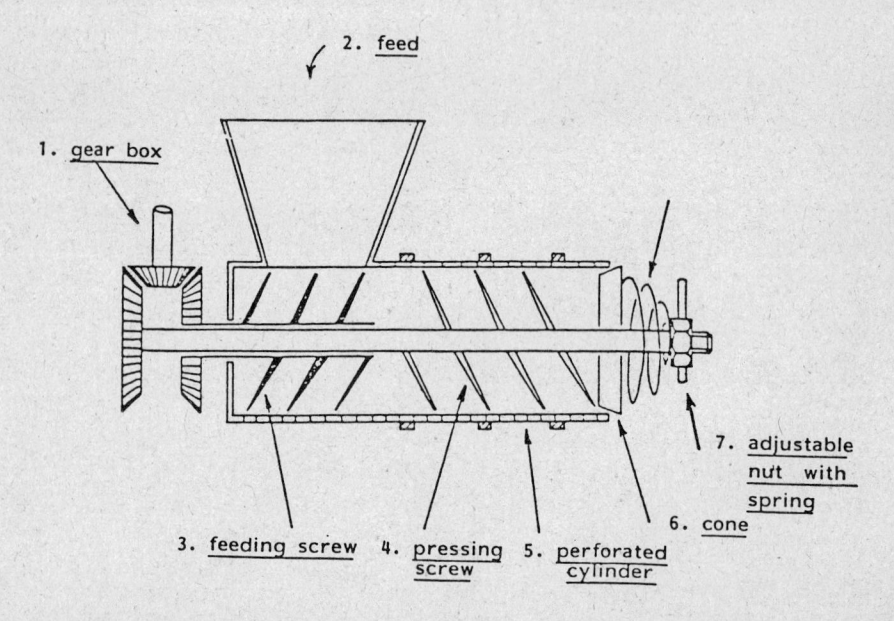

The cage is equipped with bars to prevent rotation of the cake. Cage and screw are both manufactured from wear-resistant steel. In case the screw becomes damaged by wear it can be reshaped by welding. The screw turns at 5-10 rpm, when driven by a petrol engine of 4.5 HP. The press is equipped with a gear reduction box with ratio 1/150. Empty weight of the press is 400 kg. Its processing capacity is about 300 kg cooked tenera palm fruit per hour. The price of the press (1983, c.i.f., West Africa), is about F.F. 70 000.

Oil recovery depends on rawmaterial and which process is followed.

5.2.2 Motorized processing of oil seeds

„CECOCO", Ibaraki City, Japan
CECOCO sells, as its smallest oil expeller, the „HANDER" New Type 52. The cage of the „HANDER" New Type 52 is built up lengthwise of bars. The backpressure can be varied by adjusting the axial position of the wormshaft against a stationary choke ring. The bore of the cage is 60 mm diameter. Drainage length is 150 mm. Capacity is 30-50 kg oil bearing material

Figure 35: CECOCO Press, Hander 52.
Source: CECOCO

per hour. The standard power source is a 3.7 kW, three phase electric motor. The weight of expeller and motor is 235 kg. Its performance is comparable with the MINI 40 (see Table 11).

The price of the machine is (1987, c.i.f., Amsterdam): New Type 52 + electric motor on a common base: US $ 6120. The required spare parts for two years operation (or 5000 h) are estimated at:

– Taper ring   12 pcs @ $ 30
                         = US $ 360
– Worm shaft   6 pcs @ $ 240
                         = US $ 1440
– Cage bars (set) 6 pcs @ $ 240
                         = US $ 1440
– Other (10 %)   US $ 320
  Total estimated spares US $ 3560

Simon Rosedowns, Hull, England
Simon Rosedowns are the manufacturers of the small expeller press MINI 40. Its cage, which envelops the wormshaft is made up of separately cast cage rings, spaced apart by washers. By the use of washers of the required thickness, the cage can be prepared for the processing of a particular seed. The cage has 60 mm diameter bore and 150 mm drainage length. The backpressure can be varied by adjusting the axial position of the wormshaft against a stationary choke ring. The performance as given by the manufacturer is given in Table 11. The standard power source is a 2.2 kW, three phase, electric motor. Alternatively, a 5.0 kW diesel engine can be used. The weight of the press and the motor is 250 kg.

The price of the machine is (1987, f.o.b., UK):

– NINI 40 + electric motor   = 3283
– MINI 40 + diesel engine   = 4158.

Figure 36: Simon Rosedowns, Press MINI 40. Source: Simon-Rosedowns (Crown Copyright Courtesy of the Tropical Products Institute, UK)

The required spare parts for two years operation (or 5,000 h) are estimated at:

– Choke ring   6 pcs @ 44 = £   264
– Wormshaft   4 pcs @ 315 = £ 1260
– Barrel rings (set) 3 pcs @ 217 = £ 651
– Other (10 %)   £ 225
  Total estimated spares   £ 2400

IBG Montforts + Reiners, Mönchengladbach, Federal Republic of Germany
A small expeller, called KOMET, is being manufactured by IBG. Large sized material, such as copra or palmkernels, must to be crushed into pieces of about 5 mm diameter.

The design of this expeller is quite different from the design of most other small expellers, as the screw is not a tapered compression screw.

The screw turns in a cylindrical cage, halfway foreseen with drainage holes. At the end of the cage a nozzle is mounted with a hole in the centre, that restricts the flow. To adjust the backpressure, nozzles with holes of different sizes are available. The material is conveyed into a gap between the screw and the nozzle where the material is then crushed and squeezed, and heated by the hot press head. The oil flows backward and leaks out through the holes in the press cylinder. Depending on the rawmaterial the extraction rate can reach up to 90 % of the original oil content.

The most common version is the KOMET Twin Screw Oil Expeller DD/85. The capacity depends on the type of material and may vary from 20 kg/h to 100 kg/h. Capacity and oil extraction is controlled by the rate of rotation of the screw and by selection of the appropriate temperature of the press head.

Parts subject to normal wear-and-tear are press screws, press heads, press cylinders and electric heating elements.

The required spare parts for two years operation (or 5000 h) are estimated at:
– screw   4 pcs @ DM 470 = DM 1880
– head   2 pcs @ DM 220 = DM 320
– cylinder[1]   @ DM 907 = DM   —
– heating
  element   2 pcs @ DM 180 = DM 360
– other: 10 %   = DM 270
– total   = DM 2950

The weight of this machine is about 210 kg. The standard power source is a 3 kW electric motor with stepless variable gearbox. Alternatively the unit can be driven by a diesel engine (11 kW at 3000 rpm), equipped with dynamo and clutch.

[1] only required when improperly used.

Figure 37: Monforts + Reiners, „Komet" Double Screw Expeller. Source: IBG

The price including accessories is (1987, ex-factory):
- KOMET Twin Screw Expeller
  DD 85 G + electric motor
  = DM 16 395
- KOMET Twin Screw Expeller
  DD 85 D + diesel engine
  = DM 22 470

For the crushing of big nuts, kernels or copra, a „Cutting Machine" is available. For smaller quantities the KOMET Single Screw Expeller CA 59 G with 1.1 kW flanged electric motor is available. Its capacity is 6-8 kg oil-containing material per hour. Its price is DM 4190.-. This press is not available with diesel or petrol engine. On an experimental basis, hand and pedal-driven versions are available. Recently, a new Single Screw Expeller (type SS/87 G) has been developed. This machine has half the capacity of the twin Screw Expeller DD 85 G.

Maschinenfabrik Reinartz, Neuss, Federal Republic of Germany

MRN manufactures the Screw Press AP VII as a small version of their range of large expellers designed for deep pressing. This machine combines a sturdy construction with universal usability.

The cage is built up with bars. The distance between these bars is determined by small steel distance plates. To meet the specific requirements of different oil seeds, plates of different thickness can be used. The inside diameter of the cage is 100 mm and total drainage length is 525 mm. The cage is divided into 4 sections, two with bars of 87.5 mm length and two each with a length of 175 mm. The worm consists of a typical stuffing part and a typical pressing part. The backpressure may be varied, as with larger machines, with an adjustable cone. The performance

Figure 38: Reinartz Screw Press AP VII. Source: Maschinenfabrik Reinartz

of this oil expeller, as claimed by the manufacturer, is given in Table 12.

The press can be supplied complete with a three phase electric motor of 7.5 kW or a 10 kW diesel engine. Its net weight is approximately 800 kg.

The price of the machine is (1987, ex-factory):
– Screw Press AP VII + electric motor
  = DM 29 450
– Screw Press AP VII + diesel engine
  = DM 36 050

The required spare parts for two years operation (or 5000 h) and on the basis of the use of clean seed and regular rewelding of the worm, are estimated at:
– Cone   1 pc @ 696 = DM   696
– Discharge ring
         1 pc @ 467 = DM   467
– Scrapers 12 pcs @ 189 = DM 2 268
– Worm:
stuffing part 1 pc @ 1 550 = DM 1 550
pressing part 2 pc @ 1 960 = DM 3 920

– Cage bars (set)
         1 pc @ 2 596 = DM 2 596
– Other (fixing bars + 10 %)
                      = DM 2 003
Total estimated spares   DM 13 500

A cutter for crushing large nuts, kernels and copra is available as well. Also a heating unit is available, however it requires considerable investment, including investment in a boiler unit. A smaller capacity expeller is under development.

Manufacturers in Developing Countries
In Asia and Latin America quite a number of manufacturers of oil processing machinery exist, e.g. China, India, Pakistan and Brasil. This equipment can be very appropriate to the circumstances prevailing in these countries. For example, the operation of Indian equipment in India appears not to cause difficulties, despite the fact that Indian equipment requires generally considerable operational skills as well as maintenance and

repair. In other countries, however this might cause considerable problems. It has therefore to be kept in mind, that this type of equipment might not be appropriate in countries were the required skills are not as easily available as in the country of origin.

In case Indian equipment is taken into consideration, the following addresses might be useful:

Power Gahnis and Rotary Oil Mills
(Capacity 6-30 kg/h)
– Amritsar Kolhu Factory, Queens Road, Amritsar.
– Bharat Industrial Corporation. Petit Compound, Nana Chowk, Grant Road, Bombay-400-007.
– Khadi & Village Industries Commission, Sector-2, Chandigarh.
– The REX Trading Co. P.B. 5049, Madras 600 001.
– Bhakshi Ram & Co., Miller Ganj, G.T. Road, Ludhiana.
– Bharat Industrial Corporation. Petit Compound, Nana Chowk, Grant Road, Bombay 400 007.
– Durga Dass Aggarwal & Co., Miller Ganj, G.T. Road, Ludhiana.
– Laxmi Vijay Brass & Iron Works, Pratap Nagar, Factory Area, Baroda 390 004.
– Ludhiana Expeller Industries, 471, Industrial Area-B, Ludhiana 141 003.
– Manjeet Brothers, Miller Ganj, G.T. Road, Ludhiana.
– Prashar Industries, 27 Great Nag. Road, Nagpur 9.
– The Punjab Engg. Co., Miller Ganj, G.T. Road, Ludhiana.
– The Punjab Oil Expeller Co., P.B. 8, Patel Marg., Ghaziabad.
– S.P. Eng. Corporation, 39 Factory Area, Kanpur 202 012.
– S.P. Foundries, P.B. 450, Factory Area, Kanpur 208 012.
– Sundarshan Engg. Works, G.T. Road, Miller Ganj, Ludhiana.
– Tekma Machinery Mfg. Co., P.B. 8931, Bombay 400 072.
– Tinytech Plants Pvt. Ltd., Gondal Rd., Rajkot 360 002.
– United Oil Mill Machinery & Spares Pvt. Ltd., Mathura Road, Ballabdgarh (Haryana).

6. Ongoing Research and Development Work

The development of appropriate small scale oil production technology requires a multidisciplinary approach. It requires not only activities in the technical field to improve upon processes and equipment and to realize local manufacturing of the equipment as well as establishing maintenance services, but also activities in the socio-economic field. The actual socio-economic performance of a proposed technology has to be monitored and evaluated in the practical social setting.

This kind of research and development work is, for instance, carried out by the following organizations and projects:

APICA, Douala Akwa, Cameroon
Recently APICA has presented a vertical version of its CALTECH press (capacity 70-90 kg/h). Unfortunately details are not yet available.

ATI, Washington, USA
ATI has designed a ram-type manual oil press that seems to be particularly interesting for the processing of sunflowerseed with a thin husk. The equipment is said to be manufactured by Themi Farm Implements, Arusha, Tanzania on an experimental basis.

CEPAZE, Paris, France
In Mali, CEPAZE is developing an improved processing system for village scale shea nut processing which includes mechanized grinding and separation of the oil by centrifuging. Animal drawn, pedal-operated and motorized versions are currently being tested.

ENDA, Dakar, Senegal
Besides its dissemination activities ENDA works to improve upon the existing press. The development of a nutcracker has been taken up as well.

GTZ/GATE Karité (Shea Butter) Project, Bamako, Mali
As already mentioned in section 5.1.2, the existing shea nut press is still being improved. Also work is being done to improve the preparation step in order to optimize the process.

IPI, Dar es Salaam, Tanzania
On the basis of the existing equipment for sunflowerseed processing IPI is also developing equipment for processing oilpalm fruit, palmkernels and coconuts.

Karité (Shea Butter) Project, Ministry of Agriculture, Koudougou, Burkina Faso
At Koudougou, a multidisciplinary team is evaluating a range of possible technologies to improve upon the traditional Karité processing methods.

KIT, Amsterdam, Netherlands
Besides the continued development of equipment to carry out the preparatory stages (e.g. grating equipment for coconut, and equipment for the pulping of the fruit of the Seje palm Jessenia Bataua), KIT is carrying out field tests of oil production systems. Additionally, KIT is involved in the development of dissemination methods.

NIFOR, Benin City, Nigeria
NIFOR is developing, with the assistance of UNDP, oilpalm fruit processing equipment to cater for the need of small farmers. An important component of the proposed unit is a horizontal digester capable for digesting 2 tons of sterilized fruits per 8 hours. The unit is meant to be used by cooperative societies as a centralized processing unit.

TDAU, University of Zambia, Lusaka, Zambia
On the basis of the available equipment for sunflowerseed processing, TDAU assists the company LUTANDA in Kitwe, Zambia, with the manufacturing of the equipment. Also TDAU has started to develop advisory services for assisting interested parties with the establishment of sunflowerseed processing units.

Annex

1. Guidelines for the preparation of oil fruit or oil seed processing projects

1. Determine the target group (or target persons), as for instance: women's groups, pre-cooperative or co-operative, individual entrepreneur or family enter- enterprise.

2. Clarify the organization of the target group and the responsibilities of the persons involved.

3. Clarify the needs of the target group and the way they are fulfilled at present (by selling the raw material and buying oil from elsewhere or by carrying out a traditional process).

4. Investigate the present and the future availability of the rawmaterial.

5. Define possible alternative processing systems (including the traditional or a slightly modified traditional system) for processing the available raw material.

6. Investigate the availability of the equipment, estimate the required investments and determine the possibilities for maintenance and repair.

7. Determine the market potential of the main product and the by-products.

8. Investigate possible sources of finance and the possibilities for organizational, managerial and technical support.

9. Prepare for an evaluation of alternative solutions by making feasibility calculations, in consultation with the target group.

10. Present the information to the target group and evaluate the possible alternative solutions.

2. List of abbreviations and addresses

AAB: AFRICA ASIEN BUREAU
       Schildergasse 93–101
       Postfach 100 193
       D-5000 Köln 1, Federal Republic of Germany

APICA: B.P. 5946
        Douala Akwa, Cameroon

ATDA:	Appropriate Technology Development Ass. P.O Box 311 Lucknow – 226 001, India
ATI:	Appropriate Technology International 1331 H Street, N.W. Washington, D.C. 20005, USA
CEPAZE:	18, Rue de Varenne 75007 Paris, France
CECOCO:	P.O. Box 8 Ibaraki City Osaka Pref., Japan
CONGAT:	Conseil des Organismes Non-Gouvernementaux en Activité au Togo B.P. 1857 Lomé, Togo
DMA:	Division du Machinisme Agricole, Ministère de l'Agriculture B.P. 155 Bamako, Mali
ENDA:	B.P. 3370 Dakar, Sénégal
GTZ/GATE:	German Appropriate Technology Exchange, in: Gesellschaft für Technische Zusammenarbeit (GTZ) Postfach 5180 D-6236 Eschborn 1, Federal Republic of Germany
GRET:	Groupe de Recherche et d'Echanges Technologiques 34, Rue Dumont d'Urville F-75116 Paris, France
IBE:	Institut Burkanibé de l'Energie B.P. 7047 Ouagadougou, Burkina Faso
ICCO:	Interchurch Coordinating Committee for Development Cooperation P.O. Box 151 NL-3700 AD Zeist, The Netherlands
IGB Monforts & Reiners GmbH & Co:	Postfach 20 08 53 D-4050 Mönchengladbach 2, Federal Republic of Germany

ILO:	International Labour Organization CH-1211 Geneva 22, Switzerland
IPI:	Institute of Production Innovation P.O. Box 35075 Dar Es Salaam, Tanzania
Karité Projects:	
	– Projet Karité (Ministère de l'Agriculture et de l'Elevage) B.P. 58 Koudougou, Burkina Faso
	– Projet Karité (GTZ/GATE/DMA) B.P. 100 Bamako, Mali
MRN:	Maschinenfabrik Reinartz Industriestr. 14 D-4040 Neuss, Federal Republic of Germany
KIT:	Koninklijk Instituut voor de Tropen 63 Mauritskade NL-1092 AD Amsterdam, The Netherlands
NIFOR:	Nigerian Institute for Oilpalm Research Benin City Nigeria
OPC:	Outils Pour Les Communautes B.P. 5946 Douala Akwa, Cameroon
Simon-Rosedowns Ltd.:	Connon Street GB-Hull HU2 OAD, United Kingdom
SPEICHIM:	B.P. 12 F-93140 Bondy, France
TCC:	Technology Consultancy Centre University of Science and Technology Kumasi, Ghana
TDAU:	Technology Development and Advisory Unit, University of Zambia, Lusaka Campus, P.O. Box 32379, Lusaka, Zambia

TOOL: Entrepotdok 68a/69a
NL-1018 AD Amsterdam, The Netherlands

UNATA: Unie voor Aangepaste Technologische Assistentie
G.V.D. Heuvelstraat 131
B-3140 Ramsel-Herselt, Belgium

UNIDO: United Nations Industrial Development Organization
Vienna International Centre
A-1400 Vienna, Austria

Usine de Wecker:
LUX-6868 Wecker, Luxembourg

3. Calculation of internal rate of return (IRR)[1]

The internal rate of return (IRR) is the discount rate at which the present value of cash inflows is equal to the present value of cash outflows; put another way, it is the rate at which the present value of the receipts from the project is equal to the present value of the investment, and the net present value is zero. The procedure used to calculate the IRR is the same as the one to calculate the net present value (NPV). The same kind of table can be used and, instead of discounting cash flows at a predetermined cut-off rate, several discount rates may have to be tried until the rate is found at which the NPV is zero. This rate is the IRR, and it represents the exact profitability of the project.

The calculation procedure begins with the preparation of a cash-flow table. An estimated discount rate is then used to discount the net cash flow to the present value. If the NPV is positive, a higher discount rate is applied. If the NPV is negative at this higher rate, the IRR must be between these two rates. However, if the higher discount rates still gives a positive NPV, the discount rate must be increased until the NPV becomes negative.

If the positive and negative NPVs are close to zero, a precise (the closer to zero, the more precise) and less time-consuming way to arrive at the IRR uses the following linear interpolation formula:

$$i_r = i_1 + \frac{PV\,(i_2 - i_1)}{PV + NV}$$

where i_r is the IRR, PV is the NPV (positive) at the low discount rate of i_1, and NV is the NPV (negative) at the high discount rate of i_2. The numerical values of both PV and NV used in the above formula are positive. It should be noted that i_1 and i_2 should not differ by more than one or two per cent. The above formula will not yield realistic results if the difference is too large, since the discount rate and the NPV are not related linearly.

[1] Source: UNIDO, 1978, 177f.

The IRR indicates the actual profit rate of the total investment outlay and, if required, of the equity capital. The IRR of the total investment outlay can also be used to determine the conditions of loan financing since it indicates the maximum interest rate that could be paid without creating any losses for the project proposal. In order not to endanger the liquidity of the project, it would be necessary, however, to adjust the loan repayment schedule to the cash inflows.

The investment proposal may be accepted if the IRR is greater than the cut-off rate, which is the lowest acceptable investment rate for the invested capital. If several alternatives are being compared, the project with the highest IRR should be selected if IRR is greater than the cut-off rate.

4. Summary of handoperated processes

Crop	Coconut	Groundnut	Sunflowerseed	Palmkernels	Sheanuts
Raw Material	Copra partly dried, about 30% m.c.)	Decorticated	Partly decorticated (in decorticator + winnower)	Broken kernels (in hammer mill, with 5 mm sieve)	Pounded kernels (in motar, with pestle)
Process	Grating (in disc grater) Heating (to 80 °C) and drying Pressing (max. pressure 50 bar) Drying of oil	Crushing twice (in roller mill) Moistening (~12% water) Heating (to 80 °C) and drying Pressing (max. pressure 60 bar) Drying of oil	Crushing (in roller mill) Moistening (~12% water) Heating (to 80%) and drying Pressing (max. pressure 60 bar) Drying of oil	Crushing (in roller mill) Moistening (~18% water) Heating (to 80°C) and partly drying (about 10% m.c.) Pressing (max. pressure 60 bar) Drying of oil	Moistening (if raw material dry) Heating (to 120 °C) Keeping hot (for 1–2 hours) Pressing (max. pressure 120 bar) Repeating of process Cleaning and drying of oil
KIT/UNATA UNIT:[1] Processing Oil output/day	with 14–18 persons 350 coconuts in 5 hours/day 26 kg	with 6–8 persons 80 kg/day in 6 hours/day 28 kg	with 6 men 150 kg/day in 8 hours/day with 15 persons 300 kg/day in 6 hours/day 29–58 kg	with 6–8 persons 80 kg/day in 6 hours/day 28 kg	with 5 women 40–60 kg/day in 6–9 hours/day 16–24 kg
IPI UNIT:[2] Processing Oil output/day	under development	–	with 7 men 210 kg/day in 8 hours/day 42 kg	under development	–

[1] Source: KIT
[2] Source: IPI

5. Currency conversion table[1]

US $	DM	FF	F CFA	K	£	FL	BF
Dollar	Deutsche Mark	French Franc		Kwacha	Pound Sterling	Dutch Gulden	Belgian Franc
1	1.92	6.36	318	14.36	0.67	2.17	40

[1] as of January 2nd, 1987; equivalents to US $ 1,—.

6. Literature

- African Training and Research Centre for Women: Traditional Palm Oil Processing; Women's Role and the Application of Appropriate Technology. United Nations Economic Commission for Africa, Addis Ababa, 1983.
- Agrotechnology, KIT: Oil extraction, Source XV, No. 1, pp. 102–124, 1987.
- Barrett, J. C., T. W. Hammonds and R. V. Harris, A: Technical and economic evaluation of a small scale coconut expeller operation in the Cook Islands. Coconut Research and Development, Vol. III, No. 2, 1987.
- Corbett, S.: A new oil press design: but is it any better?, in: VITA NEWS, April 1981, VITA Washington D.C., 1981.
- Donkor, P.: A hand-operated screw-press for extracting palm oil: Appropriate Technology V, No. 4. pp. 18–20, 1979.
- Eckey, A. E.: Vegetable fats and oils. New York, Reinhold Publishing Corporation, New York, 1954.
- Food and Agriculture Organization of the United Nations: Report of the first African small-scale palm oil processing workshop, NIFOR, Benin City, Nigeria 12–16 October 1981, FAO, Rome, 1982.
- Chungu A. S.: An instruction manual for IPI sunflower processing equipment, IPI, Dar es Salaam, 1986.
- Godin, V. J. and Spensley, P. C.: TPI Crop and Product Digests No. 1, Oils and Oil seeds, Tropical Products Institute, London, 1971.
- Groupe de Recherche et d'Echanges Technologiques: Le point sur l'extraction des huiles vegetales; les presses à huile. GRET, Paris, 1984.
- Hammonds, T. W., R. V. Harris and N. MacFarlane: The small-scale expelling of sunflowerseed oil in Zambia: Appropriate Technology XII, No. 1, pp. 27–28, 1985.
- Hammonds, T. W. and A. E. Smith: An industrial profile of small scale vegetable oil expelling. Tropical Development and Research Institute, Report G 202, 1987.
- Harris, R. V. and A. A. Swetman: Small scale sunflower seed processing in rural Zambia. In press (Tropical Science 1988).
- International Labour Office: Palm oil processing. Technologies for rural women – Ghana, Technical Manual No. 1. ILO, Geneva, 1985
- International Labour Office: Small-scale oil extraction from groundnuts and copra. Technical memorandum No. 5. ILO, Geneva, 1983.
- Jacobi, Carola: Palm oil processing with simple presses: GATE – questions, answers, information 4/83, pp. 32–35, 1983.
- Korthals Altes, F. W., R. Heubers and R. J. H. M. Merx: Research into small scale systems for processing agricultural products at the agrotechnology section. In: Royal Tropical Institute, Review of agricultural programmes and advisory activities 1982. KIT, Amsterdam, 1983.

- Marchés Tropicaux: Les Oléagineux, in No. 2112, 1986, p. 1167ff.
- Niess, Thomas: Angepaßte Technologie für Dorffrauen; Entwicklung von Karité-Pressen in Mali, Friedr. Vieweg & Sohn Braunschweig/Wiesbaden, 1986.
- Niess, Thomas: New shea butter technology for West African women: GATE – questions, answers, information 2/83, pp. 15–17, 1983.
- Niess, Thomas: Shea butter project provides on-the-job training. GATE – questions, answers, information 2/86, pp. 25–27, 1986.
- Rehm, S. and Espig, G.: Die Kulturpflanzen der Tropen und Subtropen, Verlag Eugen Ulmer, Stuttgart, 1984.
- Swern, E., ed.: Bailey's industrial oil and fat products. Interscience Publishers, New York, 1964.
- The Courier, European Community Publication: Dossier on Tropical Oil Seeds, in No. 86, 1984, p. 52ff, Brussels.
- Thieme J. G.: Coconut Oil Processing. FAO, Rome, 1968.
- United Nations Industrial Development Organization:
  Guidelines for the establishment and operation of vegetable oil factories. United Nations, New York, 1977.
- United Nations Industrial Development Organization:
  Information Sources on the Vegetable Oil Processing Industry. UNIDO Guides to Information Sources N. 7. United Nations, New York, 1977.
- United Nations Industrial Development Organization:
  Manual for the Preparation of Industrial Feasibility Studies, Vienna/New York, 1978.
- Weiss, E. A.: Oilseed Crops, Longman publ., London, 1983.